PIERS
OF SUSSEX

PIERS
OF SUSSEX

MARTIN EASDOWN

The
History
Press

First published 2009

The History Press
The Mill, Brimscombe Port
Stroud, Gloucestershire, GL5 2QG
www.thehistorypress.co.uk

British Library Cataloguing in Publication Data.
A catalogue record for this book is available from the British Library.

ISBN 978 0 7524 4884 8

Typesetting and origination by The History Press
Printed in Great Britain

CONTENTS

ACKNOWLEDGEMENTS

Linda Sage; The National Piers Society; Step Back in Time, Brighton; Hastings, Eastbourne and Brighton Libraries; National Archives, Kew.

Professor Cyril's Aerial Cycle Dive, West Pier, Brighton, *c.* 1910. The diver was sadly killed performing this same act on the pier two years later. (Marlinova Collection)

INTRODUCTION

Sussex has a good claim to be the birthplace of the seaside pleasure pier, for although Ryde Pier on the Isle of Wight has the earliest origins, the famous Chain Pier at Brighton was the first to be used as a fashionable promenade. Erected in 1823, principally as a landing stage for the boat service to Dieppe, the Chain Pier was immediately popularised by Brighton's wealthy visitors and residents as a health-giving walk over the sea. Little shops and refreshment outlets were opened in the towers supporting the chains, and amusements, such as band concerts and a camera obscura, were provided. The pier was visited by three monarchs and was painted by both Turner and Constable. It was, without a doubt, the place to be seen in Georgian and early Victorian Brighton.

By the late 1860s, however, the popularity of the weather-worn Chain Pier was usurped by the new West Pier which, due to its troubles of recent years, has probably become the most well-known pier of all. The West Pier in its heyday was rightly lauded as one of our finest piers, and its various stages of development epitomised the evolution of the pleasure pier. Opened in 1866 as an exclusive promenade over the sea, by the 1890s the visiting public expected more from a pier than just a bracing walk, and in 1893 a pavilion was erected at the sea-end of the pier. Competition from the rival Palace Pier, after 1899, led to the addition of a concert hall in 1916 and a range of amusements. However, the popularity of the pier declined from the 1950s and in 1975 it was closed. The pier's subsequent high-profile demise through thirty years of dereliction, fire, storms and a rescinded Lottery grant became a tragic-comic opera, made all the more poignant by the gaunt iron skeleton that remains. And yet, hope springs eternal that the pier will one day be restored, if the i360 Brighton Eye proves to be a success, as anticipated.

Brighton's third pier was the Palace Pier, completed in 1901. Now known as 'Brighton Pier', it has been a success story amongst piers, and is one of Britain's most popular free attractions. The pier manages to combine the modern (amusement rides and machines) with the past, retaining much of its ornate wrought ironwork, along with the tollhouses and signal gun from the long-lost Chain Pier.

Brighton also boasted a 'moving pier' – the extraordinary Brighton & Rottingdean Seashore Electric Railway, affectionately known as the 'Daddy-long-legs'. Conceived by Magnus Volk and engineered by the pier designer Richard St George Moore, the electric car of the line was built to resemble a section of pier to enable it to travel through the sea at high tide. The car docked at light steel landing stages at each end of the line, the one at Rottingdean resembling a traditional seaside pier. The life of the railway was short, barely five years, yet it still manages to intrigue those who view its old photographs. The Daddy-long-legs was indicative of a city that has always striven to be different.

An Edwardian scene of relaxation on Hastings Pier underneath a canopy of the pavilion. The building was unfortunately destroyed by fire in 1917. (Marlinova Collection)

The Golden Age of pier building commenced in the 1860s, and was to last for the next fifty years until 1910. In addition to Brighton West Pier, Sussex also saw two further piers erected in the 1860s, at Worthing and Bognor, and two others commenced at Hastings and Eastbourne. Worthing Pier survives in good condition, having been extensively re-modelled during the inter-war period, and it retains a period feel of that time. The same cannot be said for the poor old pier that remains at Bognor, whose once long slender neck is disappearing bit by bit. Piers which have all their amusements based in a building at a shore-end are prone to have their promenade decks neglected beyond the money-making area, and Bognor is no exception. A further shrinking of the pier in 2008 saw it lose the popular Birdman festival.

Hastings is another Sussex pier in trouble: the decay of the substructure forced the closure of the pier on the grounds it was unsafe. In 2008 the future looked glum for the pier, and we can only hope that another Brighton West situation does not develop.

However, in general, the Sussex coast has been luckier than other areas in retaining the majority of its piers. Discounting Brighton West, it has lost only two piers, Brighton Chain and St Leonards, and still retains six, including three of the finest in Brighton Palace, Eastbourne and Worthing. They will, at least, ensure that the seaside pleasure pier remains a feature of the Sussex coast for many years to come.

1

HASTINGS PIER

Hastings is a town forever linked to a defining moment in English history, when William of Normandy defeated King Harold II at the Battle of Hastings in 1066. The town then proceeded to become an important member of the Cinque Ports, guardians of the vulnerable south-east coast. However during the seventeenth century its harbour was washed away and the town settled down to become a quiet fishing port.

The town's origins as a seaside resort began during the sea-bathing craze among the wealthy in the middle of the eighteenth century. Thomas Hovenden, landlord of the Swan, was one of the first to take advantage of the trend by offering accommodation to visitors. In 1788 James Stell opened a circulating library and, six years later, produced the town's first guide. In 1824 Pelham Place was built by architect Joseph Kaye and a considerable portion of the cliff below the Norman castle had to be removed to accommodate it. Placed in the castle, at the centre of the crescent, was the church of St Mary's, built in a classical Roman style and consecrated in 1828. The development of the rival resort of St Leonards, from 1828, just to the west of Hastings, spurred further growth in the latter, aided by the coming of the railway in 1846, with a direct connection to London being completed six years later. A new town was built on the America Ground and Bohemia to the west of the old port: Patrick Robertson had leased the America Ground formerly belonging to the Crown and, by the 1860s, had developed Carlisle Parade, Robertson Terrace and Queens House. Between 1851 and 1871 the population of Hastings doubled to 29,291.

During the early 1860s the provision of a promenade pier was strongly mooted, leading to the formation of the Hastings Pier Co. in 1866, with a capital of £25,000 in £10 shares. An Act of Parliament to build a pier was granted the following year and the first meeting of shareholders, which included the town's Mayor and MP, took place at the Castle Hotel on 23 July 1869. The attractive design of Eugenius Birch for a 910ft long and 45ft wide pier[1], which included a large oriental style concert pavilion on the 125ft × 30ft pier head, was chosen. The contractor to build the pier was announced as Robert Laidlaw of Glasgow, who had submitted a tender of £23,250.

The first of the 360 cast-iron supporting piles was sunk at 3.00 p.m. on 18 December 1869, and in March 1870 the first cargo of iron was delivered to Whitstable Harbour. The heavy screw piles were then taken to Hastings and stored on the parade ready for use. By 15 July 1870 one hundred of the screw piles had been driven in.

---

[1] Earlier in the decade, Birch had drawn up plans for a harbour facing East Beach Street and the fish market, which was not built.

Above: Hastings Pier photographed soon after it was opened on August Bank Holiday 1872.

Left: An interior view of Hastings Pier Pavilion, featured on a stereograph card by F.S. Mann of 13 Wellington Place, Hastings.

Right: A programme for the evening's entertainment in Hastings Pier Pavilion on Monday 7 August 1899, which featured orchestral pieces directed by the pier's musical director, Dr Abram.

Below: The sea-end of Hastings Pier, featured on a postcard used on 18 April 1911. The children playing on the beach add colour to this evocative scene.

A postcard entitled 'Escorting Harry Lauder off the Pier at Hastings Sep. 14 1908'. Harry Lauder was one of the biggest music hall stars of his day.

Little's Entertainers photographed on Hastings Pier by A.M. Breach, *c.*1910. The Pierrot costumes were popular with concert parties before the First World War.

Above: A fine view of the ornate Hastings Pier Pavilion, *c.* 1908. The building shows the oriental influence so popular with seaside pleasure buildings. The kiosk on the left is selling postcards, whilst the other two in front of the pavilion are ticket booths for the steamers *Lady Rowena* and *Brighton Queen*.

Right: Professor Davenport was one of a number of divers who performed off Hastings Pier during the Edwardian era.

Local photographer Angus Croyle captures Professor Cyril's bicycle dive off Hastings Pier. Cyril was sadly killed while performing on Brighton West Pier in 1912.

From 1910 the deck of Hastings Pier began to be covered with additional attractions. At the entrance was placed a Joy Wheel, and beyond that there is a building showing animated pictures. The hut close to the pavilion housed a shooting gallery.

A rare view inside the Joy Wheel on Hastings Pier, *c.*1911. The purpose of the ride was to see who could stay longest on the floor in the centre as it revolved ever faster. The photograph was taken by W.J. Willmett, who was based on the pier.

In 1914 the shore-end of Hastings Pier was sold to Hastings Corporation for £7,000, who widened it to create the 'parade extension'. A classically influenced bandstand was placed on the extension. The photograph was taken by Judges, still in business in Hastings today.

The parade extension following the erection of covered seating, photographed by Judges. A new entrance to the pier can be seen under construction.

On 15 July 1917 the wonderful Hastings Pier Pavilion was destroyed by fire, thought to have been started by a discarded cigarette. This view of the fire was taken by F.J. Parsons, newspaper proprietor of Hastings, Bexhill and Folkestone.

The complete devastation of the head of Hastings Pier, as seen from an Airco machine.

An unusual postcard showing open-air dancing on Hastings Pier in 1921.

In July 1922 a rather plain pavilion was erected on Hastings Pier to replace the one destroyed by fire five years earlier.

The band enclosure and pier, c.1935. The pavilion on the pier had been remodelled to make it look more attractive. Dancing was very popular on the pier at this time.

The interior of Hastings Pier Pavilion during the 1930s.

Hastings Pier Pavilion photographed in 1982.

Hastings Pier photographed from the shore in 2000. The pier was closed to the public, awaiting its restoration by its new owner, Ian Stuart.

A view of Hastings Pier in September 2007, with the majority of the structure closed, unsafe and with an uncertain future.

The pier was officially opened on the first ever August Bank Holiday in 1872, and although the occasion took place on a day of very stormy weather, it did not stop virtually the whole town coming out to celebrate the event. The Earl of Granville, Lord Warden of the Cinque Ports, was asked to open the pier, and he travelled down from London by special train to be met by a Guard of Honour composed of the coastguard, fire brigade, artillery and rifle volunteers, the Mayor and other civic officials. Thousands of spectators lined the route to the pier in spite of the driving rain. At the pier entrance, Lady Granville and other female dignitaries were taken on to the pier in bath chairs surrounded by men with umbrellas. Lord Granville and his party toured the pier before declaring it open, and noted with satisfaction that the pier did not vibrate at all, despite the appalling weather. The celebrations then continued in the comfort of the pavilion where 600 people sat down to a grand dinner. During his after dinner speech, the Earl coined his well-used phrase, the 'Peerless Pier', when he said, 'this pier appears to me to be a peerless pier – a pier without a peer, except perhaps this unfortunate peer who is now addressing you'. The 600 guests (minus a few of the dignitaries) paid 4s each for the meal, which consisted of quarters of lamb, sirloin of beef, gelatines of veal, roast fowl, lobster, followed with pudding and blancmange, all washed down with champagne at 7s 6d per bottle. During its first week of opening, over 24,000 people paid the 2d admission charge to walk on the pier.

Aside from the pavilion, the only buildings to originally adorn the pier were two small kiosks at the entrance. A large open area had been left in the forked entrance section to enable heavy seas to pass through the decking without damaging the pier. However, during a severe gale on 1 January 1877, one of the entrance kiosks was blown over, although it was soon repaired.

The attractive Pier Pavilion could seat up to 2,000 people, and was home to its own orchestra, directed by C.H.R. Marriott. Attractions, in March 1877, included Professor Codman's *Royal Punchinello* and Signor Rudolph, the 'man flute'. In September of that year, 32,000 passed through the pier turnstiles in one week. From 1879 to 1909, when he resigned, following a disagreement with the pier directors, J.D. Hunter was entertainments manager of the pier, and he provided a broad spectrum of singers, orchestras, comedies, pantomimes, dioramas, dramatic productions and variety artists. These included the Mohawk Minstrels (who were an attraction for a number of years), Madame Haydee and her Marionettes, and Professor Day and John Alston, who were Punch and Judy performers. John Alston's last performance was at 9 p.m. when, accompanied by a pianist, he performed using light from oxy-acetylene lamps. Herr Moritz Würm and his Blue Viennese Orchestra was also a strong feature, with Gustav Holst as a member of his orchestra at one point while he was still a student of trombone at the Royal College of Music, eking out his scholarship funds. Famous names appearing at the Pier Pavilion included Zena Dare and Harry Lauder. In 1899 the pavilion was enlarged and re-roofed.

Divers were also an entertainment feature of the pier. Professor Cyril dived off the pier on a bicycle, while Augustus Davenport leaped off in a burning sack. Professor Houghton, who billed himself as the 'Human Torpedo', waged £25 he could propel his body like a torpedo for a full 35 yards and 15ins through the sea with his head fully immersed. Jules Gauiter dived off the pier with lighted fireworks strapped to his body.

Steamer trips were also available from the pier. In 1885 a landing stage was added, at a cost of £2,000, and in the same year the Hastings & St Leonards Steamship Co. was formed by A. Payne and a syndicate of local people who purchased the steamer *Carrick Castle*. In the following year they also acquired the *Albatross*, but the company was soon wound up. Another local syndicate, headed by J. Buckett, chartered the *Nelson* in 1887, whilst R.R. Collard ran the notoriously slow *May*. Mr Buckett formed the Hastings, St Leonards & Eastbourne Steam Boat Co., which passed to W.J. Butterwick in 1890. In 1888 the *Lady Brassey* replaced the *Carrick Castle*, but she was sold in 1891, having been replaced by the paddle tug *Conqueror* in 1890 and the iron paddle steamer *Seagull* in 1891. The *Glen Rosa* was also chartered in 1892 from the Victoria Steamboat Association. Plans to extend the landing stage right around the pier head were proposed by Frank Robinson in 1889, but the board of trade initially refused to sanction the work. However, in the following year, the extension to the stage was carried out.

The Brighton, Worthing & South Coast Steam Co. also ran trips to Hastings, but they folded, along with the Hastings, St Leonards & Eastbourne Steam Boat Co., in 1905, due to competition from P&A Campbell, who began running to Hastings, *c.*1890. They would continue to operate on the South Coast until 1956 when the *Glen Gower* made its final trip to France.

In 1909 a proposal was put forward to build another pavilion at the shore-end of the pier in a bid to increase the pier's income. Following successful seasons in 1906 and 1907, the pier suffered a decline in popularity in the next couple of years and a dividend was not paid to shareholders, although an offer to buy the pier from the St Leonards Pier Co. was refused. Hastings as a whole was experiencing a slump during this period (ironically when many other resorts were booming), and even its population dropped in numbers. Some blamed the decline on the increasing numbers of trippers driving away the long-staying wealthy visitors, while others cited the lack of new attractions: the town seemed unsure of what type of customer they should cater for. Hastings also suffered from a slow train service and a promenade regularly exposed to stormy seas.

In response, new attractions began to be added to the pier, including a Joy Wheel (1910–14), which revolved its customers around at an ever-increasing speed, on a glazed floor, before they were shot off into cushioned walls. Also added were a shooting gallery, slot machines, a bowling alley, rifle range and an electric theatre showing films (they were also shown in the Pier Pavilion), whilst the Hastings Pier Co. took direct control of the pier's catering in 1911. That year saw the Joy Wheel bring in £798, the Pier Pavilion £4,308 and the turnstiles £3,231. In 1912 the company achieved a net profit of £990, with Pierrot troupes bringing in £181 of income and the electric theatre £196.

In 1913 negotiations took place with Hastings Corporation with regards to them purchasing the shore-end of the pier, and the sale went through the following year at a cost of £7,000. The borough engineer, Philip H. Palmer, widened the old entrance to the pier for a 'Parade Extension', and placed upon it a bandstand with sheltered seating. During 1914 the Hastings Pier Co. gained a net profit of £875, with income being derived from tolls (£2,416), the Pier Pavilion (£1,859), the rifle saloon (£305), bowling alley (£209), steamers (£250) and the Joy Wheel (£190), which was then removed to make way for the new shore extension.

The money gained by the Hastings Pier Co. from the transaction was used to add further buildings to the pier during 1915–16, including shops, a tea room and a games room. The Aristocrat's concert party was a very popular attraction on the pier during those two years.

Unfortunately, during the afternoon of 15 July 1917 the wonderful oriental pier head pavilion was destroyed by fire. The blaze was first reported around 2 p.m. and, fanned by a strong south-westerly wind, the building was soon engulfed in flames. A huge crowd gathered on the beach and promenade to watch more than seventy firemen and troops fight the fire, although it soon became apparent that the pavilion was beyond saving. The whole building collapsed following a great explosion and the firemen were kept busy fighting the fire up until 7 p.m. The blaze was thought to have been started by a discarded cigarette end following a concert held for troops that morning.

The pier head was left a mass of twisted metal for a number of years until it was re-decked and used for open-air dancing until July 1922 when a rather plain pavilion, seating 1,400, was opened. The centre of the pier was widened from 45ft to 80ft and the Pier Theatre, seating 645, was added in 1926. By this time the pier could boast a whole range of attractions – dancing in the Pier Pavilion, concert parties in the Pier Theatre, amusement arcade, large tea room and restaurant, bowling alleys, skee-ball saloon, a bandstand on the pier head, fishing, soda fountains, box ball alleys, diving displays by Jack Hurst and steamer trips on the *Brighton Belle* along the South Coast.

In the 1930s the pier received a new art deco frontage and the pavilion was remodelled in an effort to make it more attractive. The Pier Theatre also received a makeover in 1934, and temporary buildings on the pier were replaced two years later. The pier proved to be extremely popular during this decade, with the Hastings Municipal Orchestra and Harry Hanson's Court Players attracting good crowds. The latter were to perform repertory on the pier right up until 1966.

During 1937–38 the pier's piling was strengthened. However, in 1938 the Pier Pavilion suffered serious storm damage and had to be closed while repairs were carried out, at a cost of £22,000. Two years later the pier was breached as a defence measure and the shore-end was used as a RAF training base. The breach was repaired at the war's end, and the pier was reopened in June 1946. The south-west corner of the landing stage was repaired by Messrs. Holloway Brothers.

The pier was further modified in 1951, with the opening of the West View Solarium and covered way. The East View Solarium was added five years later. In 1966 the Parade Extension saw the erection of the Triodome building to commemorate the 900th anniversary of the Battle of Hastings. The Triodome housed embroidery, made by the Royal National School of Needlework, which featured twenty-seven panels on English history since 1066. The building was a purpose-built structure of aluminium alloy, designed by Modular Metals of Maidenhead and erected by Aluminium Alloy Fabrications Ltd of Woking. In 1968 the Parade Extension was sold back to the Hastings Pier Co., who converted the Triodome to an amusement arcade. By now the pier was a popular venue for rock band concerts. Amongst the bands and acts who appeared were the Rolling Stones (four times), Jimi Hendrix, Pink Floyd, The Who, Tom Jones, P.J. Proby, 10 C.C., The Hollies, The Searchers, The Troggs and Wizzard.

Between 1872 and 1982, the Hastings Pier Co. record was tarnished by a single year, 1946, in which it ran at a loss, and that was due to the effects of war. Unfortunately, the situation was soon to change for the worse. In 1983 Hamberglow Ltd took over the Hastings Pier Co. (although they retained the name), and in November of that year was faced with a bill of £100,000 after the pier suffered storm damage. The pier began to lose money heavily, and the company struggled to keep their heads above water. On 13 October 1999 they finally succumbed to the inevitable and went into liquidation, with debts of £160,000.

The pier was closed, but in 2000 was acquired by millionaire Ian Stuart. Following repairs, the first section was reopened in May 2001 with an arcade of shops reflecting Mr Stuart's desire to create a 'Covent Garden over the Sea'. By 2003 the pier's attractions included the Pub on the Pier, run by Quay Leisure, which played host to a singles club on the first Thursday of the month. The ballroom staged tea dances on Thursday afternoons as well as student discos at weekends. Robert & Brian's Bistro served freshly prepared dishes to live piano music, and there were also two other cafes, a small amusement arcade, a bingo hall with licensed bar and a variety of retail outlets.

The pier's future seemed assured, but, sadly, it was not to be. On Friday 16 June 2006, the pier was closed on safety grounds after pieces of metal from rusted support beams were found to have fallen from its underside. It appears that the pier's restoration in 2000–01 had been merely from the deck upwards; the under-deck ironwork had been left to decay.

The pier remained closed while Ravenclaw Developments (who were running the pier, although the actual owner of it, by this time, had become very 'foggy') contested the council's use of emergency powers to shut it down in the interest of public safety. In July 2006 traders on the Parade Extension, at the shore-end of the pier, won a legal battle to keep that part open, although they had to pay for their own security and rubbish collection. However, this too was closed on 31 October 2006, and a well-attended meeting to discuss the pier's future was held in the White Rock Theatre on 12 November, which was addressed by local councillors and interested parties such as the National Piers Society. On 5 March 2007 the council agreed to commission a structural survey of the pier, at an estimated cost of £100,000.

The dispute between the council and Ravenclaw over the safety of the pier was heard at Lewes Crown Court on 27 April 2007. The court ruled that the council acted properly in issuing an emergency closure notice to Ravenclaw Investments. The company's assets were frozen and they were ordered to pay £28,000 in compensation to Stylus Sports Ltd, proprietors of the bingo club and bar on the pier, for loss of earnings and redundancy payments. The court disallowed a management buy-out of the pier following an appeal by Stylus, who thought their compensation may be at risk. Stylus paid out £200,000 for strengthening work underneath the bingo hall and Bar Luxor, enabling the central section of the pier to be reopened in July 2007. However, the pier had to be closed again during the following weekend after it was struck by lightning, and 200 people had to be evacuated.

During a hearing at Hove Crown Court on 1 November 2007, Ravenclaw Investments, and their UK representatives Boss Management, were each fined £40,000 plus costs for 'lamentably failing to meet their responsibilities under the Health & Safety at Work Act and putting public safety seriously at risk'. Furthermore, they had ignored orders issued by the council, through an improvement notice, to carry out a structural survey of the pier.

The Friends of Hastings Pier was formed in 2007 and manned a table on the reopened Parade Extension collecting signatures. It is estimated that around £17.5 million would be needed for full restoration of the pier and ongoing maintenance over the next ten years. Demolition would cost £4 million.

2

ST LEONARDS PIER

St Leonards-on-Sea was a purpose-built resort, developed by James Burton, to the west of Hastings, to which it was officially joined in 1885. Burton, born on 29 July 1861 as James Haliburton, was one of the country's top architects, and had assisted John Nash in laying out Regent Street. He first drew up his vision for the new town of St Leonards after visiting the area in 1815, and purchased the land of the Eversfield Estate (including a 1,151-yard frontage to the sea) for £7,800. The foundation stone for the first building, the St Leonards Hotel, was laid on 1 March 1828 by John Ward of Tunbridge Wells, whose father's estate at Calverley was being laid out by Burton's son Decimus. By the end of the year the South Colonnade and Boundary Arch with Hastings had been completed, and work continued on creating a whole new town boasting elegant rows of classically designed houses.

The provision of a pier for St Leonards was not part of Burton's plans, and the first application for a pier was not until 1869, when a 1,000ft structure was proposed opposite the Victoria Hotel and Baths. Nothing further was heard of this plan, and a proposal by the St Leonards Pier Co. in 1872 to enclose part of the promenade for a pier was refused by the St Leonards Commissioners. Another aborted scheme saw plans deposited for a pier by engineers Clarke & Moore, in 1884. Nevertheless, the Moore in that partnership, Richard St George Moore, was appointed engineer for a new St Leonards Pier Co. (originally the St Leonards Pier, Public Baths & Aquarium Co.), formed in September 1886 under the chairmanship of Richard James Reed. The company held their first meeting in April 1887, and were incorporated with a capital of £25,000 in 5,000 shares of £5 each. However, a loan of £7,000, at three per cent per annum, had to be obtained from the South Eastern Railway before work could start.

Five companies bid for the tender to erect the pier: Head, Wrightson; Thames Ironworks; Laidlaw & Sons; Caine, Hocking & Turtleby and Ansdell & Co. The experienced pier builders, Head, Wrightson, won the contract to build the substructure of the pier (the erection of a pavilion, bandstand and shelters was to be placed under a separate contract), and the ceremonial driving in of the first pile was carried out by the Lady Mayor on 1 March 1888. The pier was to be 960ft long and utilise over 1,500 tons of steel. An ornate Moorish-style pavilion, designed by F.H. Humphries, was to be placed at the shore-end, and a carriage drive connect it with the promenade. Additional features were a bandstand on the pier head and a small landing stage. A description of the pier was given in *The Engineer* on 11 May 1888:

> The total length of the pier is 960ft, divided into two parts, the general rule of placing the pavilion at the seaward extremity being departed from, and instead it is placed about 200ft from the shore. From the parade to the pavilion the pier is 40ft wide, having a pathway on each side

10ft wide, and a carriageway 20ft wide, so that visitors may alight from their carriages under the porch of the pavilion. From there to the octagonal head the pier is 25ft in width, with enlargements every 130ft for sheltered seats. In the centre of the head there is a bandstand with sheltered seats ranged round it.

Round three sides of the pier head there is a timber landing stage, totally independent of the pier, with landing places 6ft above high water ordinary spring tides at H.W.O.S.T. 6ft below it and 6ft above L.W.O.S.T.; this landing stage is built of 13 by 13 pitch pine piles strutted and braced with iron.

The screw piles are 12in in diameter, and 11in in thickness. The blades vary from 3ft to 2ft 6in in diameter, and they are screwed in according to the strata from 10ft to 16ft. The piles are collected in groups from four to eight in number, strongly braced and strutted together; each group, therefore, forming independent and self-supporting piers. The strutting is formed of rails 56lb per yard, and the tie rods are 1¾in in diameter, secured with patent lock nuts. No cast iron lugs are used, the bracing being attached to the columns by straps passing round them between two collars, cast on to prevent them slipping up or down. The columns have capitals, secured by studs riveted over, underneath the girder bed-plates. The girders are lattice girders throughout, bolted together over the piles. The main joists are rolled steel, 5ft centre to centre; on these there are secondary joists, 2ft centre to centre, carrying the flooring, which is laid across the pier with ⅛ open joists. The main body of the pavilion joins a concert hall, to seat about 600 people. The west wing is devoted to two large subscription rooms, a drawing room, and smoke room. At the east end there is a public refreshment room. Round the building there is a colonnade, cast iron columns with cast iron arches. The main girders for the roof are semi-circular lattice girders, supported on cast iron columns, embodied in the wall of the building. The total estimated cost is between £19,000 and £20,000.

The pier substructure took two years to construct, and on 9 October 1890 the Mayor and his wife, along with Councillor Stubbs, attended a ceremony to fix the last bolt. However, it was to take another year before the pavilion and the other on-deck buildings were completed. The final cost of construction was £30,000, and the pier was officially opened by Lady Brassey on 28 October 1891.

The Pier Pavilion was also known as the Kursaal, and could seat 750 patrons. The building additionally featured a restaurant, tea room, ladies drawing room, smoking room and billiards room. Until the First World War, the Kursaal boasted its own pier orchestra.

A description of the pier and pavilion appeared in the *Hastings and St. Leonards Observer* on 31 October 1891:

The style adopted in the building of the pier, which is 950 feet long, is somewhat unusual, the pavilion being at the shore-end, and approached by means of a carriage drive, so that in rough weather vehicles can drive right up to the very doors. At the far end of the structure there is a large extension, upon which it is intended to build a band kiosk, and in addition to this, the sides of the main pier will be relieved with handsome alcoves. Quite at the extremity also a landing stage has already been erected, and at the lowest spring tide there will be six feet of water there, so that in almost any weather pleasure steam boats will be able to run alongside.

ST. LEONARDS-ON-SEA PIER COMPANY,

LIMITED.

Incorporated under the Companies' Acts, 1862 to 1883.

Nominal Capital £25,000, in 5,000 Shares of £5 each.

No. 643   £5 SHARE

This is to Certify that *John Bray*
of *13 South Colonnade St Leonards*
is the holder of the Share numbered 643 in the above-named Company,
subject to the Articles of Association thereof, and that the full sum
of **£5** has been paid up on the said Share.

Given under the Common Seal of the Company

this 31st day of *August* 1888

R. J. Reed

William Carbers

} Directors.

H N Gould

Secretary.

This Certificate must be produced before the Registration of any Transfer of Shares can be made.

A share certificate of the St Leonards-on-Sea Pier Company, issued to John Bray of 13 South Colonnade, St Leonards, on 31 August 1888.

St Leonards Pier nearing completion in 1891.

St Leonards Pier photographed by G.W. Stokes during the Edwardian period. The landing stage at the end of the pier was damaged by storms in the 1890s and was not repaired.

During a gale on 26–27 November 1905 a shop at the entrance to St Leonards Pier was overturned and wrecked. Judges have captured an inquisitive crowd viewing the disaster.

The orchestra based at the St Leonards Pier Kursaal, photographed in 1904.

A fine postcard by Angus Croyle showing the St Leonards Angling Festival of 27 October 1907.
The photograph was taken from the upper storey of the pavilion.

In 1909 St Leonards Pier was acquired by an American syndicate, who re-styled it the 'American Palace Pier' and added a skating rink at the end of the pier.

On the American Palace Pier, St Leonards–on–Sea, 1909. The bandstand was moved from the pier head, and features a performance by a group of Pierrots. The kiosks and skating rink were newly added that year.

Another view by Judges taken on St Leonards Pier in 1909. Keeping in with the American theme, one of the kiosks is selling ice cream sodas.

Inside the skating rink on St Leonards Pier in 1909.

A poster highlighting the attractions on St Leonards Pier, which include roller-skating, an American bowling alley, dancing, angling, concert parties and refreshment rooms and bars.

The interior of St Leonards Pier Pavilion.

A postcard showing the Musique Minicipale De Rouen on St Leonards Pier on 8 July 1912.

St Leonards Pier in the mid-1930s, sporting the gaudy frontage added by the Paget Guarantee Corporation, who acquired the pier in 1933.

The fatally damaged St Leonards Pier pictured at the end of the Second World War.

A plaque was unveiled in 1996 on the site of St Leonards Pier commemorating the showing of early moving pictures in the pavilion.

The under structure differs somewhat from the Hastings Pier, not only in its spans, which are almost double the length of those of Hastings, but also in the position of the piles. These are placed in such a manner that each is totally independent of the others. The piles are made in three pieces, which rest in another, and the lower piece is fitted with a very broad flange, which has been screwed into the clay foundation by hydraulic power. The ornamental capitals were screwed on after the columns were up, and out of the 1,500 tons of iron used not one ounce was condemned, a fact which speaks well for the contractors, Messrs Head, Wrightson and Co., of Stockton-on-Tees.

At the entrance the tollhouses are finished off in a remarkably smart manner with neat roofs and light pinnacles, while the iron work on the larger structure is both light and elegant in appearance. Before this building and at the south side, below the steps, are erected a pair of really magnificent lamps, surmounted by electric light, and at the end of the deck a third one is placed.

The interior of the pavilion gives one the impression of a semi-Moorish style of architecture, with its light columns and frieze of floral iron work, the whole being picked out in terra-cotta. The orchestra is constructed on the east side, and like all the fashionable theatres, etc, the drapery of green and gold, and curtain of maroon velvet, caught up in graceful folds from the centre to the sides, gives this part a very handsome appearance. By day light is admitted on all sides through coloured and frosted glass, while at night gas and electricity are used. Seating accommodation is provided for 750 people, and the club rooms at the south sides, each 36ft by 20ft, take up space that can be thrown into the larger hall.

The landing stage, however, proved to be little used and, in October 1896, it was wrecked in a storm. Portions of the washed-away timber smashed into Hastings Pier, causing damage to the ironwork. The stage was reconstructed the following year, but suffered further storm damage in 1899. Although boats were no longer unable to call, the stage was used by fishermen and divers until it was dismantled in 1909. A further storm in 1905 toppled over one of the tollhouses at the pier entrance.

The heavy expenditure on maintenance and repairs to the pier added to the woes of the St Leonards Pier Co., which had been in debt from the start, having owed the contractors £20,000. On 1 April 1909 the pier was acquired by an American syndicate and restyled the American Palace Pier. New tiered seating was installed in the pavilion and the orchestra pit was lowered in front of the stage. Electric lighting was installed all along the pier, and six ornamental kiosks were added. The bandstand was moved to up behind the pavilion to make way for a roller-skating rink on the pier head. Attractions on the pier in 1911 included the Pier Orchestra, variety shows, open air fetes, confetti carnivals, an American bowling alley and the skating rink. Angling remained popular and many competitions were held on the pier.

Further improvements were carried out in the early 1920s. In October 1921 a covered corridor was erected from the pier gates to the pavilion and, in March 1923, a new bowling alley was opened. That same year Julian H. Clifford was appointed musical director of the pier. His appointment was announced in the *Daily Telegraph* on 11 August 1923, which emphasised the rivalry that still existed between Hastings and St Leonards:

Julian H. Clifford has been appointed musical director to the St Leonards Pier Co. by Mr J.H. Gardiner, the managing director, and founder of the Julian Clifford scholarship. Lest any confusion should arise as to the possible connection between St Leonards, with Julian H.

Clifford, and Hastings, next door, it should be noted that no such connection exists, despite the contiguity of the two resorts.

However, the pier was still losing money, and the substructure needed £21,000 in repairs. In November 1927 Hastings borough council refused an offer to purchase the pier for £9,000. Eventually, in August 1933, it was sold to the Paget Guarantee Corporation, fronted by the brothers David and Philip Lannon, and Charles Genese, who acquired the debentures of the old company. The Lannons, who were known to the police for running dubious financial enterprises, brought in a receiver and appointed their brother-in-law, Arthur Collins, to run the 'New Palace Pier' as lessee of 'Southern Piers Limited', enabling them to keep in the financial background as far as the registered particulars were concerned.

Upon their acquisition of the pier, the Lannons announced that the skating rink was to be removed and the clubroom for the St Leonards Sea Anglers converted into a soda fountain. There were also plans for a new landing stage and open-air dancing platform. What the people of St Leonards did get, much to the horror of some of them, was a gaudy new art deco frontage, and entertainments such as wrestling, a zoo and performing fleas. There was even a funeral laid on for one of the fleas in the act, who it was said had been killed after being struck by lightning! The funeral proceedings included a band playing a funeral march, a funeral cortege walking along the pier with wreaths donated by stallholders, the last post, two minutes silence and the casting of the body into the deep!

The Lannons also employed a man in a motorboat to speed along the coast shouting through a loud hailer what the pier had to offer. However, this attracted the attention of the police, as did the rowdy licensed bar and the Lannons' habit of giving away scores of complimentary pier tickets. In November 1934 a protest by the rector of St Leonards Church led to the Sunday dancing on the pier being ordered to be stopped. At the same time, ambitious and unrealised plans were announced to spend £120,000 on an extension to the pier featuring a swimming pool and jetty.

On 28 September 1938 the Lannons decided to pull out of the pier and put it up for auction. The pier was described as having a deck area of 62,500ft and a ballroom accommodating 1,000 people with a maple sprung floor, café, sun lounge, bars and shops. However, it failed to reach its reserve price. In the following year, Southern Piers Ltd proposed adding an arrow-shaped landing stage with two berths to replace the existing stage with its three small berths. However, the coming of war meant that the work was not carried out.

The pier was breached as a defence measure in June 1940, and suffered bomb damage later that year on 4 October. The sad end for this fine pier came on 7 March 1944 when a fire destroyed the pavilion. The blaze was allegedly started by a Canadian soldier at around 5.40 p.m., and eight fire pumps fought the fire for over an hour.

The pier was now just a charred wreck and beyond any hope of effective restoration. Nevertheless, this did not prevent the council announcing, on 20 November 1945, that they would control any future development of the pier. Needless to say the remains of the structure were left as an eyesore and in December 1950 the council were forced to acquire it for demolition. The sea-end of the pier crashed into the water during a storm in March 1951, and the remains of the shore-end were removed shortly after. A plaque was erected at the site of the pier in 1996 commemorating the showing of early moving pictures in the Pier Pavilion.

3

EASTBOURNE PIER

As the popularity of sea bathing took off in the middle of the eighteenth century, visitors began to frequent the small fishing hamlet of Sea Houses, south of the village of Bourne. Accommodation was erected and the settlement of 'East Bourne' created. In 1787 James Royer produced a guide to the area, which was visited by Princess Amelia in 1789 and 1790. By 1795 a circulating library, billiard rooms and baths had been established, along with a theatre in nearby South Bourne and the Lamb Assembly Rooms. However, the growth of the resort remained small until 1858 when William, 2nd Earl of Burlington, became the 7th Duke of Devonshire, the local landowner. Building upon the arrival of the railway in 1849, the Duke began co-ordinating a high-class development of the town, encouraging developers by offering financial assistance of up to seventy-five per cent in advances on the required capital sum. The first section of Marine Parade had been laid by architect James Berry in 1847, and the building of Grand Parade commenced in 1851. The Duke appointed Henry Currey as his agent, and in 1859 they produced a plan for the layout of the new resort. Despite an early setback, when a number of the properties failed to find a buyer causing developers to sell back to the Duke who then had to dispose of them at auction, the town grew rapidly into an aristocratic resort favoured by the wealthy.

In 1864 the Eastbourne Pier Co. was formed under the leadership of Lord Edward Cavendish, the Duke of Devonshire, with a capital of £15,000, and an Act of Parliament for the construction of a pier was granted the following year. The original position of the pier was opposite Devonshire Place, but this was soon moved to near Cavendish Place. The land upon which the shore-end of the pier was to stand was leased from the council for 999 years, at five shillings a year. The simple design of Eugenius Birch was chosen for a 1,000ft-long iron pier with 60ft-span girders, supported by 25ft-long cast-iron columns 12in in diameter and screwed 7ft into the seabed. The iron piles were to sit in special cups in the rock bed to enable movement in bad weather. The pier neck was 22ft wide, except where it had two pairs of 86ft recesses, and boasted continuous seating along both sides. Two tollhouses were placed at the entrance to the pier, and four further kiosks were placed on the pier head along with a windbreak. The contract for building the pier was awarded to J.E. Dowson, although his death, in 1868, led to Head, Wrightson taking over the contract.

The first pile was driven in on 18 April 1866 by the Marquess of Hartington, but progress proved slow, and by March 1870 only three spans were in position. The *Eastbourne Standard* reported:

> The total length of the pier will be about 1,000 feet and will terminate with a spacious head having landing stages on either side so that steamers can land passengers at any state of the tide.

Since the death of J.E. Dowson, Esq, the former contractor, the works have been carried out by Messrs. Head, Wrightson, & Co., the present contractors, under the superintendence of Mr Henry Matravers, the resident engineer, and the whole of the ironwork has been cast at their Teesdale Iron Works at Stockton-on-Tees. In March last there were only about three spans of girders erected, but since that period four more spans have been added. The pier now runs out a distance of 500 feet. The pier has a clear deck width of 22 feet, with comfortable seats on either side, the whole of its length being relieved by two recesses 68 feet wide. The body of the pier will consist of twelve bays or spans of girders 60 feet long, which are supported by cast iron columns let into very strong screw piles that penetrate into the bed of the sea a distance of 7 feet. The bed consists of very hard blue clay. The columns are 12 inches in diameter and 25 feet in length, fixed into the screw piles eight feet long. The second recess will, when finished, form a convenient space for refreshment stalls, and the band during the summer months will play to the delight of the visitors to Eastbourne. The termination of the seventh span was tested on Sunday last by the firing of two six-pounders, and it was found that there was not the slightest vibration thus showing that the Pier is amply strong to resist any seas that may come in contact with it.

Although it was far from finished, it was decided to open the first stage of the pier with great ceremony on 13 June 1870. A procession through the town to the pier was headed by the County Police and the Eastbourne Town Band. The Duke of Devonshire was accompanied by the directors of the Eastbourne Pier Co., and in the carriages behind were the engineers and contractors. The fire brigade, Oddfellows and Foresters, then followed and bringing up the rear was the lifeboat and its crew, on a transport carriage drawn by six horses. Following the opening ceremony, a dinner was held at the Assembly Rooms, and there was a display of athletic sports on the cricket field.

The pier was not fully completed to a length of 1,000ft until 1872, having cost a total of £13,400 to build. The staff employed on the pier consisted of a Pier Master (the first was John Vine), one deck hand and the tollhouse keeper. Admission onto the pier was 2d, and further revenue was gained by the sale of advertising spaces at the gates and along the façade of the deck. Bands were engaged to perform in the pier head bandstand, although Sunday concerts were eventually discontinued, and, even as late as 1926, no kiosks on the pier were allowed to open on a Sunday. In the mid-1870s the bandstand was moved to the slightly more sheltered centre of the pier. For two seasons, in 1872–3, Thomas Gowland, a director of the Eastbourne Pier Co., operated the paddle tug *Rapid* from the pier. Mr Gowland additionally sold newspapers from one of the pier's kiosks.

The *Eastbourne Standard*'s predictions that the pier 'was amply strong to resist any seas' proved to be sadly wrong when the deck was weakened by rough seas washing over it. Birch had designed the pier too 'low', and works had to be carried out to strengthen it. However, this proved to be of no avail for, during a severe storm on 1 January 1877, the shore-end of the pier was completely washed away. The Pier Master, Mr Sawdie, and two employees named Henry Barber and Caesar Mitchell, had a lucky escape from the collapsing pier by jumping for safety, although Barber did suffer a broken leg.

The shore section of the pier was rebuilt to a higher level than the original sea-end, and a slope connected the two (which is still noticeable today).

Right: A view along Eastbourne Pier, *c.*1872. Taken from a stereograph card.

Below: A carte de visite of Eastbourne Pier, *c.*1875. The two hexagonal kiosks on the pier were added a few years after it was opened.

On 1 January 1877 the shore-end of Eastbourne Pier was completely wrecked during a severe gale.

Following the 1877 storm, the shore-end of the pier was rebuilt to a higher level than the unaffected sea-end. The bandstand had been moved to the centre of the pier from the pier head in the mid-1870s.

In 1888 a rather plain pavilion was added to the pier. Accommodating 400 people, it cost just £250.

A view of the pier showing the pavilion and the join between the two sections of pier, c. 1890.

The Pier Eastbourne

FCC Series 29

Eastbourne Pier Co, Ltd

New Pavilion.

Manager —— Mr. G. Hayes.

Programme.

ALL the CLOCKS about this Pavilion are supplied by
WM. BRUFORD & SON, The Clock and Watch Makers,
EASTBOURNE & EXETER.

Above: In 1901 the pier was remodelled with a fine new pavilion, which featured a 1,100-seat theatre, tea room, upper balcony and, in the dome, a camera obscura.

Left: A programme for Eastbourne Pier Theatre for the week commencing Monday 19 August 1901, advertising a performance of *Lady of Ostend* by Mr Lawrence Brough's company.

A postcard view of the pier issued by the French postcard publisher L.L., *c.*1907. The ornate bandstand and small pavilion beyond were added as part of the 1901 improvements. The pavilion was one of two added where the deck sloped from the shore- to sea-end, housing a rifle range and bowling saloon.

A postcard advertising a performance of The Follies in the Pier Theatre, who gave two performances per day.

The entrance to Eastbourne Pier photographed on a snowy night in March 1909.

A wonderfully busy view of the pier in 1910, with an early motor bus adding to the scene. The original entrance kiosks were replaced in 1912, although the centre kiosk still survives and can be seen close to the Pavilion Tea Room.

During the roller-skating craze of 1909–10 an area of the pier deck around the bandstand was set aside as a rink. This postcard was published by Judges of Hastings.

The pier bandstand and theatre in 1909. The windscreen was made of cast iron decorated with dolphin motifs, and was added in 1903.

Eastbourne Pier with the new entrance kiosks added in 1912.

Captain Kettle's 70ft 'English Header' dive off Eastbourne Pier in 1913.

PIER ENTRANCE AND NEW PAVILION, EASTBOURNE.

Eastbourne Pier in the mid-1920s showing the new Music Pavilion built in 1925.

EASTBOURNE PIER BALLROOM AND RONNIE HANCOX'S B...   9268

An interior view of the Music Pavilion/Ballroom with an inset of Ronnie Hancox's Band. This postcard was posted on 5 August 1955.

The remodelled pier entrance added in 1951.

The interior of
Eastbourne Pier
Theatre is featured
on the front cover of
a programme for the
1954 season of the
musical *Star Wagon*.
The show featured
future *Carry On* star
Peter Butterworth.

SANDY POWELL

Sandy Powell and his show *Starlight* was a feature of the Pier Theatre from the 1950s to 1969.

The theatre on Eastbourne Pier sadly ceased to be used following damage sustained in an arson attack in January 1970.

A look along Eastbourne Pier in 1982.

In 1991 the pier entrance was revamped for a third time with this Victorian-styled addition.

In 1888 the pier acquired its first pavilion, a plain shed-like structure accommodating 400 people, at a cost of only £250. Five years later a new three-berth landing stage, designed by Frank Robinson and constructed by Thomas Gibson from the hardwoods jarrah and greenheart, was added. To assist with the cost, the Eastbourne Pier Co. requested a loan of £2,000 from the Duke of Devonshire. The stage was extended in 1912.

Further improvements were carried out to the pier in 1901. The pavilion was carted away to be used as a shed by a farmer at Lewes, and a far more impressive structure, designed by Noel Ridley, replaced it. The new theatre had two-level seating for 1,100 people, a large tea room on the balcony facing the landing stage, licensed bar and offices. The theatre was primarily used for first class plays and entertainments, although the first engagement was Wolfe's German band, who received £3 for performing four (later six) times a day, seven days a week. Shows were presented all year round and, to make life more comfortable for winter patrons, heating was added in 1906. The Pier Theatre was the first venue in Eastbourne to show animated pictures.

A camera obscura was installed in the dome of the new theatre: one of the largest in the country, it revolved on huge ball bearings turned by a hand windlass. The silver-surfaced mirror and lens were mounted at an angle of forty-five degrees, and projected the image outside on to a 6ft-wide bowl painted with emulsion.

Two small pavilions in matching style were additionally added to the centre of the pier, which housed a rifle range and American bowling saloon. A bandstand was also erected (which remained in use until 1939), and in 1903 heavy cast-iron windscreens, decorated with dolphins, were laid along the pier. The improvements of 1901–03 had transformed the pier from a rather ugly duckling into a most elegant structure. Larger entrance kiosks replaced the originals in 1912.

During the First World War, troops from the Summerdown Convalescent Camp performed their plays on the pier. In 1915 husband and wife Felgate King and Elsie Mayfair produced their summer show *Pier Revels*, which ran until 1934. The pier's attractions during the 1920s can be gleaned from the 1924 edition of the *Ward Lock Guide*:

> strong swimmers can take a header from the Pier Head any morning between 6–9 a.m. fee 4d; season tickets 6s. There is also a large dressing room for bathers, with a special approach from the landing stage. In the large pavilion, concerts and dramatic entertainments are given. There are also tea rooms, a camera obscura, an American bowling saloon, a roller skating rink and a rifle range. Admission 2d, 1 dozen tickets 1s 9d, annual ticket 10s 6d.

The camera obscura was run, before the Second World War, by Mr Pelly, who was also an agent for P&A Campbell which, in 1908, was running the vessels *Brighton Queen*, *Cambria*, *Glen Rosa* and *Bonnie Doon* from the pier.

A further attraction was added to the pier in 1925 with the opening of the Pier Ballroom at a cost of £15,000. Admission onto the pier was raised at the same time to 3d. The plans for the ballroom had been submitted in 1923 but had been initially rejected by the council on the grounds that they would obstruct the sea view of nearby properties. However, following amendments, the plans were passed. Seating 900, the ballroom boasted a girder-slung rock maple floor and the latest lighting effects.

Clarkson Rose's *Twinkle* became a feature of the ballroom, and it was this comedy that was in progress when, in May 1940, at around 10.10 p.m., a party of sappers arrived to blow a hole in the pier for defence purposes. The officer was persuaded to delay the act, and the staff were given three days to remove their valuables. The delay proved to be a godsend as it was eventually decided to remove just the decking rather than breach the pier, as was the case with most South and East Coast piers. A bofors gun was eventually placed at the shoreward-end of the pier, and the camera obscura was home to a machine gun. The pier suffered damage when a mine exploded at the shore-end, damaging the wall of the ballroom and pushing the building 2ft to one side, while machine-gun fire holed a 10,000-gallon water tank, causing water to flood the offices below.

Following the end of the war, the pier was reopened in August 1946 after most of the decking was rebuilt in concrete. A new concrete landing stage was also added, and, in 1951, the entrance was remodelled. The bandstand had been removed, but the camera obscura was still a feature, and the Pier Games Saloons offered, amongst others, 'The Skee-Roll', 'Derby Greyhound Racer', 'Bisley Rifle Range' and Darts. P&A Campbell offered sailings from the pier to Brighton, Hastings, Worthing, Folkestone and the Isle of Wight, or, alternatively, there were the motor vessels *Matapan* and *Dunkirk*. The 'Gay Musical' *Star Wagon*, starring Sandy Powell, was the main feature in the theatre, whilst there was dancing nightly to Ronnie Hancox and his Dance Band in the ballroom. Admission to the pier was raised to 4d in 1954 and 6d in 1961.

The 1950s were profitable times for the Eastbourne Pier Co. and, due to a high reserve of capital, a one for one bonus share was issued, raising the share capital to £80,000. Yearly dividends for the years 1947–60 were: 1947–50 – seven and a half per cent; 1951–52 – five per cent; 1953–54 – six per cent; 1955–56 – seven and a half per cent; 1957–58 – eight per cent; 1959 – nine per cent and 1960 – eight per cent.

Changing leisure tastes led to the ballroom being converted into an amusement arcade in 1968. The new arcade, termed the 'Blue Room', featured automated amusements, children's rides, bowling, licensed and coffee bars and a small jive bar. The Pier Theatre remained home to the popular 'Mr Eastbourne', Sandy Powell, in his show *Starlight*, for many years until, in January 1970, the interior of the building was sadly destroyed by a fire started by a pier employee.

In 1968 the pier had been acquired by Trust House Forte for £170,000. Their successors, First Leisure, gave the pier a £250,000 facelift in 1985, although the Great Storm of 16 October 1987 badly damaged the landing stage. The pier's attractions in 1988 included the Roxy discotheque, the Channel Bar (featuring long-time organist and entertainer Chris Mannion), the Blue Room amusement centre, Professor Peabody's Play Place and Ma Beeton's restaurant. During the winter of 1990–91 a new entrance in Victorian style was built by WFC of Devon at a cost of £500,000, which featured two octagonal pavilions connected by a clerestory roof upon which was a turreted clock. A further addition to the pier's attractions was added in 1995 with the inaugural Birdman competition, mirroring the contest already held off Bognor Regis Pier. There was also a Round the Pier raft race. In 1997 the Eastbourne Pier was the National Piers Society's 'Pier of the Year'.

On 30 May 2003 the pier's camera obscura was reopened. Situated in the dome above what had become the Atlantis Club, it had been inaccessible to the public since the pavilion fire had destroyed the access ways. With financial help from the Conservation Area Partnership Scheme, a new double stairway was erected, at the top of which a spiral staircase led up to the dome.

Having passed into the hands of Six Piers Ltd, in February 2006, the Grade II listed pier is a handsome structure that still manages to attract people of all ages.

4

THE BRIGHTON & ROTTINGDEAN SEASHORE ELECTRIC RAILWAY

The short-lived Brighton & Rottingdean Seashore Electric Railway must have presented quite an amazing spectacle even during those late Victorian days of engineering excellence. For many it seemed as though a piece of seaside pier had broken away and was moving by itself through the sea!

Affectionately known as the 'Daddy-long-legs', 'spider car' or 'sea car', the railway, as well as housing a mode of transport resembling a section of pier, had a couple of other pier connections. The engineer was Richard St George Moore, the designer of the impressive St Leonards and Brighton Palace piers, who used his pier-building experience in the design of the car. The other pier connection was the building of two light steel landing stages to enable passengers to board and land from the sea car. One of them, at Rottingdean, somewhat assumed the role of a traditional seaside pier by being used for promenading and fishing.

The brains behind the Daddy-long-legs was the German clockmaker Magnus Volk, who had opened the first electric railway in the country at Brighton on 4 August 1883. This originally ran from the Aquarium to the Chain Pier, with later extensions to Banjo Groyne (1884) and Black Rock (1901). Happily this railway survives to this day and stands as a fine memorial to the pioneering German, who also installed Brighton's first telephone system, electricity in the Royal Pavilion and a street fire alarm system, which is still used all over the world.

First proposed in 1892, an Act of Parliament for the extension of the Volk's Railway from Paston Place (Banjo Groyne) to Rottingdean, through the sea, was gained on 27 July 1893, and construction was commenced the following year. However, progress was slow because the sea submerged the line for much of the time. The first part of the project to open was the pier at Rottingdean, on 11 June 1895; a light steel structure of 300ft length and 20ft width situated just to the west of Rottingdean Gap, to which it was connected by a short walkway[2]. The pier stood 30ft clear of high water, and steps led down from the 60ft-high pier head to the landing stage. Beneath the head was located the railway's generator of electricity, a 100hp steam generator. This provided generated electricity to an overhead cable, supported by poles laid along the length of the line. This cable, being in contact with two small wheels held up

[2] The original 1892 proposal for the pier at Rottingdean showed it to be 220ft in length with four sets of supporting columns. The right-angled entrance slope from the pier to the gap was also missing.

against it at the top of the conducting rods, upon the car, continually gave off an electric current, which passed through the 24ft-high pier-like legs to the driving wheels and brakes. The car ran upon two parallel lines of rails, each line forming a single railway of 2ft 8½in gauge, which was laid apart to a space of 18ft. Concrete slabs laid on the seabed supported the rails, which were placed 60–100 yards away from the cliffs.

The 50 × 22ft platform of the car consisted of an upper open deck with seating, and a luxury saloon (25 × 12ft) on the lower tier. This was fitted with a fine upholstered Ottoman down its centre, stained-glass windows, carpet, potted plants and aspidistras, heavy curtains and a refreshment booth. The builder was the Gloucester Wagon Co., and the car was officially named *Pioneer*.

The Board of Trade inspection of the completed line noticed a number of faults which needed to be remedied before it could opened. These included an incomplete sleeper formation, omitted nail clips, the fixing of cast-iron sleeper chains, a buffer stop at Rottingdean and, as regards the car, the addition of life buoys, fire buckets, a boat, small companion ladder and a railing around the driver at each end of the vehicle. Once these were carried out, a licence was issued on condition that no more than 150 passengers were carried in the car at any one time. Furthermore, there was to be a maximum speed limit of 8mph, no running in bad weather and a daily inspection of the track. The staff were to wear naval-type uniforms.

Eventually, on 28 November 1896, after costing £30,000 to build, Magnus Volk opened his new wonder railway through the sea with an official ceremony at the Brighton terminus. Present at the opening ceremony were the Mayor of Brighton, Alderman Blaker, and the Chairman of Rottingdean parish council, Mr Steyning Beard, who were accompanied by members of Brighton town council and other invited guests. The assemblage of dignitaries then proceeded to board *Pioneer* (except the local MP, who had another engagement to attend) for the two mile, seven furlong, one chain, thirty-five-minute journey to Rottingdean. A celebration lunch was also held at the Madeira Hall Shelter.

The public service was started on 30 November 1896, and, not surprisingly, the railway was initially a great success, with large crowds flocking to the Brighton terminus at Paston Place (adjoining the Banjo Groyne) to patiently await their turn to experience a ride through the sea. However, disaster was to strike the railway on 4 December 1896 when a severe gale wrecked both the Paston Place terminus and *Pioneer*, after it broke loose from its mooring at Rottingdean Pier. Additionally, the Chain Pier was completely destroyed, and timbers from the wrecked structure damaged the West Pier.

Undeterred, Volk rebuilt *Pioneer* (with the legs 2ft higher), and built a new smaller landing stage off the Banjo Groyne at the Brighton end, enabling the service to be resumed on 20 July 1897. A request stop was opened at Ovingdean Greenway Gap using a rather fragile sloping wooden landing stage 220ft long and 40ft wide.

On 20 February 1898, the Prince of Wales took two trips on the railway. Indeed, even one ride on *Pioneer* was really only for those wealthy enough to afford the 6d fare each way. An hourly service was provided to Rottingdean in the summer, although short trips from Banjo Groyne became increasingly popular. This was due to a journey along the full length of the line and back, taking up to one and a half hours to complete, at an average speed of only 6mph. At high tide the car would crawl at a walking pace, and for some the journey could become quite tedious. Breakdowns were also common, causing the timetable to be suspended for weeks on end, whilst bad weather also brought the service to a halt.

However, there was no denying a journey on *Pioneer* was a unique and unusual experience, especially when it was ploughing through a high tide of 15ft of water. The *Morning Post* described it as:

> A railway which carried its passengers over the surface of the sea to afford all the pleasurable sensation of a yachting trip without fear of 'Mal de Mer' is a thing so strange and unlike every engineering enterprise yet conceived that it merits more than a passing notice.

Unfortunately, the life of the Brighton & Rottingdean Seashore Electric Railway was to be all too brief. During July and August 1900 the service had to be suspended when the track was damaged by the scour from the construction of two concrete groynes to help prevent the erosion of the cliffs. A number of Brighton residents also complained that they thought the track was generally unsafe. Then, on 1 September 1900, Brighton Corporation gave Volk two months notice to relocate the track at Kemp Town, 200ft southwards, to make room for groyne extensions. Volk suggested building a new terminus at Black Rock (and extending his shore railway to it), but this was rejected and, in February 1901, the borough surveyor removed the track and the railway was closed. In the following year St George Moore drew up a new deviation line further out to sea, complete with a new landing stage at Brighton. A viaduct along the cliffs was also suggested as a replacement, but heavy financial costs and the prospect of not recouping them meant both ideas were soon shelved. However, Volk was allowed to extend his shore railway to Black Rock in compensation.

The elegant *Pioneer* was tied up at the Ovingdean landing stage, where it was left to die a slow death along with the landing stages. Volk claimed he had no finance to maintain them or even to pay the £30 outstanding rent due to the board of trade. The shore-end of the Ovingdean Pier was removed to deter trespassers and, during 1906–07, J.J. Clark of Goldstone Farm, Hove, who was a director of the company, removed some of the rails near Ovingdean. By February 1910 Clark had removed *Pioneer* and Ovingdean Pier, save for a few stumps.

As regards the more substantial Rottingdean Pier, Clark stated that he was prepared to pay the outstanding rent and take it over. On 30 October 1908 he wrote to the board of trade:

> This jetty is in good condition and could be used, as it formerly was, as a promenade by the Rottingdean inhabitants and visitors on payment of a small toll. It might be lengthened, under a Provisional Order, so as to admit of its being used as a landing place for pleasure steamers.

The Board of Trade granted powers for the pier to be used as a promenade and for tolls to be levied, but the pier remained closed until, in 1911, Magnus Volk reported to the Board of Trade that the structure was too dilapidated. On 30 June that year the Crown resumed possession of the foreshore where the pier stood, and Volk engaged Messrs. Blackmore, Gould & Co. (who had offered £40 scrap value) to remove the pier. Work commenced on 11 September 1911, and had been completed by early December.

Left: A works photograph by the Gloucester Wagon Company of *Pioneer'* the electric car that ran through the sea between Brighton and Rottingdean.

Below: The opening day of the Brighton & Rottingdean Seashore Electric Railway on 28 November 1896.

Pioneer at the Paston Place terminus in November 1896. The landing stage was destroyed on 4 December 1896, during the storm that wrecked the Chain Pier.

Magnus Volk is pictured seated inside the plush surroundings of *Pioneer*.

The destruction of the Paston Place terminus of the Brighton & Rottingdean Seashore Electric Railway in the gale of 4 March 1896.

Above: Pioneer at Rottingdean Pier. Note the railway's electric generator under the pier head.

Right: A colourful and evocative advertisement for the 'Sea Voyage on Wheels'. The fare was *6d* each way.

Opposite below: The replacement Paston Place terminus fixed to the Banjo Groyne. The sea car, or Daddy-long-legs as it was affectionately known, departs with a full load. Note the lifeboat and lifebelts on *Pioneer.*

The Daddy-long-legs at low tide, clearly showing the track and the poles supporting the overhead electric wire.

Rottingdean Pier photographed in 1910. The pier, which had not been used by the Daddy-long-legs since 1900, was demolished the following year.

Rottingdean Cliffs and Sea-going Car.

Near Brighton

The landing stage at Ovingdean Gap. Following the closure of the railway in 1901, *Pioneer* was tied up here until it was broken up along with the pier in 1910.

A final view of the unforgettable Brighton & Rottingdean Seashore Electric Railway.

The rails at the Rottingdean end had been removed by Blackmore, Gould & Co. in December 1910, but Rottingdean parish council remained unhappy that the concrete blocks remained in place. In 1913 they requested that they should be removed as they were a danger to craft using the new slipway. Some of the blocks were removed at Rottingdean, but elsewhere along the line they remained in place – and still do so to this day, leaving, at low tide, a visible reminder of the line's existence. Stumps of some of the wooden poles that supported the overhead wire also survive. Models of *Pioneer* can be seen at both Brighton and Rottingdean museums.

Interestingly, a similar passenger platform on stilts, known as the Le Pont Roulant (the moving bridge) crossed the harbour at St Malo, using cable traction, until 1915.

So Volk's imaginative railway through the sea was, in the end, probably a failure. However, its uniqueness has ensured it has an honoured and treasured place in the annals of British seaside history.

5

BRIGHTON ROYAL CHAIN PIER

One of the most famous of all piers, and probably the first to be used for pleasure, the Chain Pier at Brighton is also the only pleasure pier to have been visited by three monarchs – George IV, William IV and Victoria. Although initially built as a landing stage for the Dieppe packets, this service soon fell away, and the pier quickly became extremely popular as a fashionable promenade.

The transformation of the fishing village of Brighthelmston into one of the country's most popular seaside towns was led by Dr Richard Russell from nearby Lewes. In 1750 he published the Latin treatise *De Tabe Glandulars*, which was translated into English and published in 1753 as *A Dissertation Concerning the Use of Sea Water in Diseases of the Glands*. This document paved the way for the future popularity of the English seaside resort by advocating the drinking of and bathing in seawater for the benefit of one's health. His theories were avidly taken up in his local seaside town of Brighthelmston (the name Brighton was not formally adopted until 1810), where Russell built himself a residence in 1754 (Russell House, now the site of the Albion Hotel), and opened the Chalybeate Spring at St Ann's Well Gardens, Hove.

Dr Russell's philosophies (and those of his successors, Dr Anthony Relan and Dr John Awister) brought the wealthy flocking to Brighton. One of the first was the Countess of Huntingdon, in 1755, who sold some of her jewels to raise money to build a small chapel behind her house in North Street. Dr Johnson was a frequent visitor, and often bathed in the sea, although he typically grumbled about the place, exclaiming that it was 'so truly desolate that if one had a mind to hang oneself for desperation on being obliged to live there, it would be difficult to find a tree on which to fasten the rope'. The first Royal to visit the burgeoning town was the Duke of Gloucester, the younger brother of George III, in July 1765. The Duke of York followed suit in 1766 and the Duke of Cumberland in 1771. The latter became particularly enamoured of the area and was instrumental in bringing the Prince of Wales to the town on 7 September 1783. The Prince stayed with the Duke at Grove House, and undertook bathing in the sea in a bid to remedy the swellings in his neck (which he hid by wearing high collars). He so enjoyed himself that he resolved to build his own residence in Brighton and, in 1786–87, engaged Henry Holland to erect the original Royal Pavilion. This was added to by P.F. Robinson, using a Chinese style, in 1801–04, before being transformed by John Nash, between 1815 and 1822, into the lavish Oriental/Indian-influenced palace that we know today.

The Steine was established as a fashionable promenade, and the Old Ship and Castle were amongst the earliest hotels, which also took on a function as entertainment centres, with their assembly room and ballroom respectively. A number of circulating libraries were established and the Theatre Royal was opened in 1807. Horse racing on the Downs first took place in August

1783, and short-lived public gardens such as the Promenade Grove and the Royal Brighton Gardens briefly flourished. To make bathing more comfortable, indoor baths were opened.

Grand housing was developed on both the western and eastern end of the seafront. Between 1798 and 1807 Royal Crescent was erected on the eastern side by J.B. Otto, and the houses were faced with black mathematical tiles. Otto also erected a buff-coloured plaster statue of the Prince of the Wales to a design by Rossi. The statue stood 7ft tall, upon an 11ft-high pedestal, and showed the Prince dressed in the uniform of a colonel of the Tenth Hussars. However, if Otto thought that the statue would put him in good stead with the Prince, he was sadly wrong, for the Prince hated it! Furthermore, the sea air eroded the structure and parts began to break off. After one of the arms fell off the statue was mistaken for Lord Nelson! It was pulled down in 1819 and, nine years later, a more substantial bronze replica of the prince was erected on The Steine (later moved, in 1922, to near the Royal Pavilion).

The peak period for growth for Brighton was between 1818 and 1828. In the western part of the town Bedford Square was built in 1818, and Regency Square ten years later. Just across the boundary, in Hove, Augustus Busby, Amon Wilds and Amon Henry Wilds erected the grand Brunswick Town estate. They were also responsible for the equally impressive Kemp Town estate for Thomas Kemp.

By 1820 Brighton was firmly established as a fashionable marine resort, and its population and visitor numbers were rising fast. The demand for passage to the Continent was ever-increasing but without a landing stage in the town the Dieppe packets had to be anchored off the beach and the passengers ferried to and from them in small boats. This demand for a safe landing pier led to the formation of the Brighthelmston Suspension Pier Co. Ltd in February 1822, with a capital of £27,000 in shares of £100 each. They confidently (and inaccurately!) estimated that yearly revenue from the pier would amount to around £8,000, of which £3,650 would be gathered from the £10 received each day from those using the pier as a promenade, whilst £2,500 was expected from the estimated 25,000 people travelling to and from France.

Captain Samuel Brown was engaged to design and engineer the pier. Brown had spent seventeen years in the Royal Navy and in 1816 had patented an improved form of links for chain cables, which the navy introduced the following year. He also improved the chain link for suspension bridges, which he put to good use on the Chain Pier. Prior to the construction of his pier at Brighton, Brown had erected a suspension bridge across the River Tweed at Berwick in 1819–20, and the small chain pier at Newhaven, Edinburgh, in 1820–21.

The Brighton Chain Pier was to be 1,134ft long, with a narrow walkway only 13–14ft wide, and at a height of 14ft above high water. Four pyramid-shaped cast-iron towers, 25ft high and spaced 260ft apart, carried the suspension chains, which were anchored to a 3ft iron plate (weighing a ton) driven 54ft into the cliff face, and beyond the pier head into the sea. The towers rested on seventy piles driven 10ft into the rock, whilst the pier head rested on 150 piles to support the 80ft platform of Purbeck stone, which weighed 200 tons. The piles were of Norwegian fir thickly coated in pitch, with metal points for easier insertion into the rock. Part of the original design for the pier included a proposed breakwater, estimated cost £3,000, which was soon abandoned[3].

3 An 1830 proposal to convert the pier into a breakwater, by filling in the outer piles with chalk and stone and lengthening the head 20ft to the south and 30ft westward, was also never realised.

Brightbelmston Suspension Pier.

111 POUNDS.

$\mathfrak{B}$**rightbelmston Suspension Pier Company,** constituted by Act of Parliament, the Third Year of our Sovereign George the Fourth, intituled " An Act for erecting and maintaining a Chain Pier, and other Works connected therewith, at the Town of Brighthelmston, in the County of Sussex."

We, the Brightbelmston Suspension Pier Company, incorporated by virtue of an Act of Parliament, do hereby certify that ZACH. KEMP, Esqr., of is entitled to THREE SHARES of the Capital Stock of this Company, being Nos. 104, 105, 106, as witness our Common Seal affixed hereto, this 18th day of August, 1823.

Entered pursuant to the Act.

FREDK. COOPER,
Clerk to the Company.

(Seal of the Company.)

A share certificate issued by Brighthelmston Suspension Pier Co. to Zach Kemp on 18 August 1823.

The Chain Pier in the year of its opening, 1823. The gates of the private entrance to the pier can be seen by the water-wheel.

The entrance to the pier in 1824 with the tollhouse seen on the right.

J. Aumonier's painting of the Chain Pier featuring the signal gun (which can still be seen on the Palace Pier), small shops in the towers, and where the chains passed through the Purbeck stone to be anchored to the seabed.

Above: The head of the Chain Pier with a brick–built camera obscura, seen on the left. Also seen is Whittlemore's Juvenile Library, housed in one of the towers.

Right: Annual admission tickets to the Chain Pier along with a share certificate.

SUBSCRIBER'S TICKET (circa 1830)

ROYAL TICKET—Used by Residents at the Pavilion.

THE STORM OF OCTOBER 15th, 1833.

SECTION OF PIER—DAMAGE TO SECOND BRIDGE.

SECTION OF PIER DAMAGE TO THIRD BRIDGE.

Left: On 15 October 1833 the Chain Pier was badly damaged by a storm that wrecked the second and third bridges.

Below: A view along the undulating deck of the Chain Pier (*c.*1880) showing a clear view of the suspension chains.

A photograph taken from the head of the Chain Pier looking towards the shore, *c.*1885. A small derrick is visible on the left, and a shop is operating in the tower on the right.

The Royal Chain Pier bazaar and camera obscura, which were sited by the entrance to the pier.

Left: An unusual view of the deck of the Chain Pier, *c.*1885.

Below: The attractive kiosks at the entrance to the Chain Pier were added in 1871, after the Aquarium entrance and private esplanade were sold.

The Royal Chain Pier photographed around 1885. The board by the machines is advertising the playing of the town band on the pier.

A photograph of the entrance to the Chain Pier showing the chains suspended across the esplanade. They were anchored into the cliff on the right.

A rare photograph showing the period between 1891 and 1896 when Brighton had three piers, although the Palace Pier in the centre was still under construction and was not opened until 1899.

The Royal Chain Pier on its last legs, photographed in November 1896. The structure had been declared unsafe a month earlier and closed.

The sad end for the Chain Pier came on the night of 4 December 1896 when it almost completely washed away during a storm. Only the first tower and the entrance kiosks remained standing.

Mr Edward Fogden (centre), shipwright and watchman on the pier for forty years, poses in front of the remains of his beloved Chain Pier.

The pier was to be originally placed at the foot of the Old Steine, but objections from local boat owners and fishermen that the pier would obstruct their landings on the beach led to the site of the pier being moved 270m east. The pier entrance, however, remained at the original site and a private esplanade, 1,250ft long and 33ft wide, led to the pier proper. However, the esplanade was sold to the Aquarium for £8,600 in 1871, and the entrance to the pier was relocated to where it met the shore. Two tollhouses were added, and these still survive today as shop kiosks on the Palace Pier.

Work on the pier began in September 1822, with Brown himself subscribing to half the shares. The Duke of Wellington was amongst the other subscribers. During the construction work, one man died and another lost his arm as a result of a falling chain. Just a month after the work began, on 13 October, a stormy sea swept away the pile driving equipment, and gales were to be a cause of frequent interruptions to the work. The original contractor, McIntosh, terminated his contract due to the difficulty of driving the piles into the seabed. Brown finished the work using direct labour (principally ex-naval men), and the final cost of construction was £30,000.

Hopes that George IV might officially open the pier were disappointed, and both the Duke of Clarence and Duke of York had previous engagements. The opening date was fixed for 23 November 1823, when Captain Brown presided over a gathering of 250 dignitaries at the Old Ship Hotel, before they left the hotel in procession at 1 p.m. led by the Town Beadles and a military band ahead of the Town Commissioners marching 'two by two, with badges and scarves'. They reached the pier at 2 p.m. to a crescendo of cheers and the playing of the National Anthem. Captain Brown presided over a luncheon in a temporary marquee on the pier head, where the toasts of the King and the Royal family were warmly received before Mr Colpoys sang a special song composed by Mr Heathcote of the theatre:

But of all the sweet pleasures Brighton can boast,
A walk upon the Chain Pier delighteth me most,
That elegant structure, light airy and free,
Like a work of enchantment hangs over the sea.

Outside, music provided by two bands entertained the spectators, and the sea around the pier was so covered with pleasure boats that 'not a vacant spot could the eye discover, but thousands of objects displaying beauty, fashion and gratification'.

At 4.30 p.m. Mr Slade, the secretary of the pier company, officially opened the pier in front of around 5,000 people gathered on the pier, and an estimated 25,000 on Marine Parade. The proclamation of the pier's opening by the town crier was made both on the pier and at the entrance gate along the esplanade. Brighton's master of ceremonies, Mr W. Forth, bought the first admission ticket to go on the pier. In the evening darkness, the crowd that remained enjoyed a celebratory display of fireworks from the end of the pier, accompanied by background music from the Bands of the 7th Hussars and 58th Regiment of Infantry. Samuel Brown concluded the day's festivities by entertaining the directors of the pier company at his home with supper and dancing in the ballroom. The floor of the ballroom sported a chalk representation of the Chain Pier, executed by the artist Edward Fox of Ship Street. In February 1824 bronze and silver medals were struck to commemorate the erection of the pier, featuring an image of the structure on one side and George IV on the other.

The pier immediately became a fashionable promenade for Brighton's wealthy visitors and residents, and on two Sundays in July 1824 some 3,000 visitors each day enjoyed a stroll upon it. Admission was 2d although an annual subscription could be bought for one guinea. The toll was usually collected by a partially deaf employee named Ratty who, when asked a question, would always reply 'tuppence please'. In addition to a bracing promenade, the pier also offered floating plunge baths, a camera obscura (moved from the beach by Russell House), shower and douche baths, bench seating, occasional divers, firework displays, the Brighton Town Band and regimental bands. The towers housed little shops that offered refreshments and souvenirs: the shop in the first tower was run by the Gates family for around forty years, and the one on the pier head was occupied by the family of Ellen Terry for around forty-four years. The Terry family became noted for their excellent cherry brandy.

On the esplanade by the entrance to the pier were three cottages with regency-style balconies through which the chains passed. The west cottage was home to the Pier Master, whilst the other two were incorporated into the Royal Chain Pier Bazaar, which also contained a library, saloon, telescopes and a camera obscura on the roof.

The first steam packet to use the pier was the *Rapid*, in May 1824, which was greeted by large crowds on the pier and the firing of cannon. The vessel sailed to Dieppe three times a week (taking nine to ten hours each way), and proved so popular that an additional vessel named *Union* was put on the route. In the following year a steamer service ran from the pier to the Isle of Wight.

Brighton's Royal patron George IV approved of the pier and visited it in December 1823, although he was reluctant to walk upon it. When the King returned to Brighton later that month, the pier was illuminated with 16,000 lamps of different colours. The first Royals to fully use the pier however were the Duke and Duchess of Clarence when they sailed from Dieppe to Brighton aboard the Royal Yacht on 15 October 1829. They were met at the pier by a large crowd and renditions of the National Anthem and Rule Britannia from the band of the escort frigate *Hyperion*. The couple returned to the pier the following year as King and Queen, and the 'Sailor King' became fond of promenading up and down it as the structure reminded him of a deck of a ship. Queen Victoria visited the pier on 20 October 1837 and was accompanied during a walk upon it by Samuel Brown. She returned to the pier six years later on 7 September 1843 in the company of Prince Albert following a visit to France. They were welcomed at the pier by music from the Grenadier Guards and a welcoming party that included Samuel Brown before being escorted to the Royal Pavilion. The pier's Royal patronage allowed it to hold the title of Royal Chain Pier.

The pier's immediate success allowed a dividend of seven per cent to be paid to shareholders in 1825. However, the shareholders were never to be rewarded as fully again as storm damage repairs began to eat into the pier company's profits. During a severe storm on 24 November 1824, the so-called 'Birthday Storm', the pier suffered minor damage whilst the sea wall nearby was destroyed. The pier suffered more fully during a storm on 15 October 1833, when it was struck by lightning, leaving a gap in the third bridge of 30-40ft. Such was the public's affection for the pier that a house-to-house collection raised £1,300; enough to repair it.

Unfortunately, the pier was damaged again during a fierce storm on 29 November 1836 that caused much damage in the town. The platform of the third bridge was lifted several feet into the air before falling into the sea after the suspension chains broke. Two men had a narrow escape as they had just previously crawled across the bridge before it collapsed. Another collection to assist towards repairs was organised, but this time the public were not so forthcoming with their donations towards the £1,000 repair bill. Brighton was experiencing a slump in fortunes at this time, which was not reversed until the arrival of the railway in 1841.

On 13 March 1852, Captain Brown (who had been knighted in 1838) passed away. The heyday of his beloved pier (which had enamoured and been captured on canvas by artists such as Turner and Constable) was already passing, and its popularity began to decline further from the 1860s as larger and more exciting attractions such as the West Pier (1866) and the Aquarium (1872) were opened. Various proposals were put forward to revitalise the pier. In 1883 the West Pier's engineer, Eugenius Birch, designed a Kursaal for the head of the pier, but he died before the plan could be implemented. Eventually, in 1889, the pier was acquired by the Brighton Marine Palace & Pier Co. The company had been formed in 1886 and was incorporated by an Act of Parliament two years later to raise £150,000 by subscription. The company was compelled to demolish the old Chain Pier and build a new pier in its place. Originally this was to be on the same site, but, when work began on what became the Palace Pier in 1891, the Chain Pier was still standing and the Palace Pier was sprouting out by the entrance to the Aquarium (on the site where the Chain Pier was originally to have been placed). The shareholders in the old Brighthelmston Suspension Pier Co. Ltd received £13 6s 8d in cash for each £100 share, and £36 13s 4d in debentures of the new company.

An examination of the Chain Pier following storm damage in October 1896 found the head to be 6ft 9in out of plumb, and the pier was closed as unsafe. Tenders to dismantle it were invited, but before this could be carried out the weather finally put the old pier out of its misery. During a severe gale on 4 December 1896 Miss Body, in the Chain Pier Lodge, felt the pier shudder violently its last death throe before witnessing it collapse into the sea. The last person who had set foot on the pier was Edward Fogden, who lit the warning lamp on the pier head. He had been associated with the pier for forty years and he too witnessed the demise of his old friend.

All that remained standing of the pier was the entrance kiosks, a few timber piles and the first clump of piles with its tower leaning over, broken and defeated. The timber from the dead pier was washed ashore all along the beach, after taking a section of the West Pier with it. A few days later, 150 lots of the timber were sold off at prices ranging from 2s 6d to 5s each. Most of the buyers were firewood dealers, although some of the timber became Chain Pier memorial souvenirs. The first clump of piles and the tower proved to be quite difficult to remove, and were not auctioned off until 2 February 1897.

Remains of the Chain Pier's clumps of piles could clearly be seen on the beach until the 1930s, and at extreme low tides they can still be seen today. The pier's old sundial was placed on the Palace Pier by the theatre, but was later removed to Balcombe churchyard. However, the pier's entrance kiosks and signal cannon can still be seen on the Palace Pier. This famous old pier may be long gone, but it is certainly not forgotten.

6

BRIGHTON PALACE PIER

The Brighton Marine Palace & Pier Co. was formed in 1886 to acquire the undertakings of the Brighthelmston Suspension Pier Co., owners of the Chain Pier. They proposed to demolished the Chain Pier and erect a new pier complete with pavilion and landing stage, engineered by Edward Wilson. An Act of Parliament was gained in 1888, and the company's capital was announced as £150,000 (15,000 × £10 shares), with authorisation to borrow a further £50,000. The Chain Pier was purchased for £15,000 in 1889, but the decision was then taken to move the site of the new Palace Pier to the foot of the Old Steine (by the site of the original entrance to the Chain Pier). This left the Chain Pier still standing and open to the public, although the Act of Parliament stated that on completion of the new pier the Chain Pier had to be demolished.

A new design for the Palace Pier was prepared by engineer Richard St George, and the first pile was driven in on 7 November 1891. The pier was to be built with cast-iron screw piles in groups of six, which supported lattice girders on which were mounted rolled steel joists. The decking was to be composed of the hardwoods keruing and kapur.

A further Act of Parliament, passed in 1893, allowed three years for the completion of the pier and removal of the Chain Pier (by 7 August 1896). An agreement was reached on 23 August 1894, with Messrs E. van Hoegaerden & Co. of Antwerp to supply steel joists, iron fish plates and other materials for the construction of the pier, in return for shares in the company. However, work stopped on the pier in March 1895, having reached a length of 1,060ft, due to exhaustion of funds, and a new Act had to be gained in 1896 to extend the building work for a further three years. The company were plunged further into crisis when, during a storm on 4 December 1896, the Chain Pier was washed away. The wrecked timbers of the old pier caused £6,000 of damage to the West Pier and £1,500 to the Volks Railway, and the company was sued for damages. In addition, the unfinished Palace Pier suffered harm to the tune of £2,000, causing one correspondent to note: 'it was perhaps a little working of the hand of fate that the Chain Pier in its last moments should strike a blow of some sort at the newer rival, which indirectly in some measure had been the cause of its abandonment.'

The company's liquidation was ordered by the High Court, but fortunately the philanthropist James Howard came to their rescue, and contractors Mayoh & Haley were able to re-start the work. A 3ft 6in-gauge works tramway was laid along the line to assist with the construction, and although permission was granted to convert it to passenger carrying, it was not retained upon the full opening of the pier. On 20 May 1899 the Lady Mayor of Brighton, Mrs Hawkes, opened the first 1,500ft of the pier to the public and laid the foundation pile for the Pier Theatre. Work on erecting the pier

On 20 May 1899 the Lady Mayor of Brighton, Mrs Hawkes, officially opened the Palace Pier and fixed the first column of the theatre.

A rare view taken (*c.*1900) of the works tramway laid along the Palace Pier to assist in its construction, but which was not retained upon the full opening of the pier in 1901. A band concert can be seen in progress, but there are not many watching it!

THE PALACE PIER THEATRE, BRIGHTON.

The Palace Pier Theatre opened on 3 April 1901. The oriental domes of the building mirrored the nearby Royal Pavilion.

The Pavilion, Palace Pier, Brighton

The original interior of the Palace Pier Theatre (c. 1904), which was extensively remodelled in 1911.

Brighton Palace Pier seen from the beach at low tide in 1907.

An interesting view of the entrance to the Palace Pier, *c.*1906. The initials of the Brighton Marine Palace & Pier Co. can be seen on the centre arch, and the tollhouses are advertising the pier toll as 2*d*. The Pier Theatre is playing host to *One Summer's Day*.

Above: The Palace Pier lights up the night sky with its 3,500 lamps, *c.*1907.

Right: An Edwardian programme for the Palace Pier Theatre featuring a classical concert by Lieutenant Dan Godfrey. The programme is also advertising sailings by the *Worthing Belle* to Worthing, Newhaven, and for an English Channel trip.

A postcard used on 17 May 1910 which shows the central windscreen added to the Palace Pier in 1906.

In 1910–11 the winter garden, designed by C.E. Clayton, was added to the Palace Pier.

The interior of the winter garden in 1911 looking towards the fountain and the band area beyond.

H.G. Amers and his band photographed by Donovan's Studio in 1913. They were a very popular attraction on the Palace Pier at this time.

Professor Reddish's bicycle dive off Brighton Palace Pier, *c.*1912.

A postcard entitled 'Brighton Air Race, the winner over the Palace Pier'. The card was posted on 11 May 1911 with the message: 'This is the winner "Hamel" driving a monoplane. He was waving his hand as he passed by.'

The Anglo–American bicycle polo team on the Palace Pier in 1907. The poster on the left advertises that all seats are free.

The Palace Pier Follies in 1923. The postcard describes them as Royal Entertainers whose patron was HRH Prince of Wales.

A 1920s view of the Palace Pier. The winter garden has been converted into the Palace of Fun, and open-air dancing is an attraction.

In 1930 the entrance to the Palace Pier was set back 40ft and a clock tower was added. This postcard was sent on 8 August 1933.

A 1936 programme for the Palace Pier, which describes itself as 'The Finest in the World'. Captain W.J. Dunn was directing the Pier Orchestra.

The interior of the Palace Pier Theatre in the 1950s; much altered since it was built in 1901.

An aerial view of the Palace Pier in the 1960s. The landing stage lies disused, and would be removed in 1973.

On 19 October 1973 a drifting barge used in the demolition of the landing stage badly damaged the north-west corner of the pier head, leaving the theatre hanging precariously over the edge.

The Palace Pier's rifle range photographed in 1975. Four years earlier it had featured in the film *Carry On At Your Convenience*.

The exterior of the Palace Pier Theatre in 1975. Externally the building remains little changed from its opening in 1901.

A 1982 photograph of one of the Chain Pier's toll houses on the Palace Pier.

The amusement dome that replaced the Palace Pier Theatre in 1986.

The aftermath of the fire on 4 February 2003 which destroyed several amusement rides on the Palace Pier.

Now known as 'Brighton Pier', the Palace Pier is one of Britain's most visited free attractions.

head theatre took nearly two years and it was opened on 3 April 1901 with concerts by the Pavilion Orchestra and the Sacred Harmonic Society. Admission to the pier was originally 3*d*, although theatre patrons were exempt from paying it.

The completed pier was 1,760ft in length and 45ft wide along its neck, covering an area of two and a half acres, and, in the end, had cost an enormous £137,000 to build. The deck was gaily decorated with arches forged at the Phoenix Ironworks in Lewes, which were lit by over 3,500 electric lamps installed by Mr M. Fileman. A central windscreen was added down the centre of the neck in 1906.

The Pier Theatre could seat 1,500 people and featured tiered seating, as well as refreshment, reading and smoking rooms. With its minaret domes, the building was clearly based on the oriental theme of the Royal Pavilion. The theatre was put to a multi-purpose use and housed a wide variety of entertainments, including Gilbert & Sullivan Operettas, musical comedies, West End farces, ballet, military bands and music hall variety shows. A covered stage behind the theatre on the pier head was home to Pierrot and concert party troupes such as the Palace Pier Entertainers. For those who took their music more seriously, there was the Brighton Municipal Orchestra, and in 1906 Sir Henry Wood, famous for his Promenade Concerts, performed a series of chamber music works. Pandering to popular taste, films began to be shown in the theatre from 1911.

The pier's landing stage could be used at all states of the tide, and steamers such as the *Brighton Queen*, *Worthing Belle*, *Devonia* and *Glen Rosa* operated trips to other South Coast resorts and to France. The diver Professor Reddish entertained the crowds, who were also charmed by a variety of 'What the Butler Saw' machines and sweetmeat vending machines. The pier had bathing rooms opened from 6 a.m. to 6 p.m. daily, and a gymnasium. The pier's central location on the seafront quickly led to it becoming a popular attraction.

During 1910–11 a number of improvements were carried out to the pier. Richard St George Moore widened a section of the pier and the winter garden, designed by C.E. Clayton, and an enclosed band area, were erected. They became home to the popular bandmaster H.G. Amers, who conducted military bands such as the String Band of the Northumberland Hussars during performances of a 'Humorous selection of Melodious Morceaux and Intermezzos and Solos for Cello, Violin, Cornet and Vocal Recitals'. The interior of the theatre was remodelled with a circle, boxes and an enlarged stage, and had a new seating capacity of 1,300. On top of the building a rooftop garden and café were added, and two lifts were provided to carry people up from the pier deck. The high teas cost 6*d* per person. The Palace Pier Restaurant was sited above the theatre entrance. A high-class establishment, it served luncheons beginning at the price of 1*s* 6*d* and Table d'Hote dinners at 2*s* 6*d* and 3*s* 6*d*. The meals were complimented by fine wines and Havana cigars and the serenading of the Imperial Ladies Orchestra. Under the pier was the Cave Café run by the Styles family, who later ran a bowling alley and fruit and vegetable stall on the pier. The landing stage was also remodelled on its north-east side with a berth for yachts.

A full programme of events on the pier on Tuesday 5 September 1911 was as follows:

09.00 a.m.	Pier opens to the general public
10.10 a.m.	*Brighton Queen* leaves for Eastbourne, Hastings and Folkestone
11.30 a.m.	Morning Performance of the String Band of the Northumberland Hussars. Vocalist – Mr W. Topliss Green

11.45 a.m.	Morning Performers of the Palace Pier Entertainers at the Pier Head
12.15 p.m.	Iced drinks of all descriptions may be obtained at the Buffet
01.00 p.m.	Special Luncheons in Restaurant from 1/6
02.00 p.m.	Fine view of Channel obtained from Pier Head. Comfortable deck chairs
03.00 p.m.	*Glen Rosa* leaves for Eastbourne
03.15 p.m.	Afternoon Performance of the Military Band of the Northumberland Hussars
03.30 p.m.	Afternoon Performance by the Palace Pier Entertainers at the Head of the Pier
04.30 p.m.	Teas served on Roof Gardens. Lifts from deck
07.00 p.m.	Dinner at Restaurant. Table d'Hote from 3/6. Roof Gardens for refreshments and wines.
08.00 p.m.	Evening Performance of the Military Band of the Northumberland Hussars.
08.00 p.m.	Performance in the Theatre of *Dare Devil Dorothy*
08.15 p.m.	Evening Performance of the Palace Pier Entertainers at the Head of the Pier
11.00 p.m.	Pier closes

Following the end of the First World War the pier continued to grow in popularity, targeting the masses, while the West Pier continued to be more select. Jack Hurst continued the diving tradition off the pier, performing between 1919 and 1921, and another popular attraction was Nelson Lees' wooden clockwork-driven models of haunted houses, churchyards and executions. The pier staged its own carnivals, which featured Bathing Belle competitions, beauty and novelty contests, and yearly pantomimes. Dance bands, such as Jack Hylton's, drew large crowds, as did the popular Brighton Follies, led by Jimmy Hunter and Jack Shepherd's Brighton Entertainers, where Max Miller began his career. The comedian of the Brighton Follies, Tommy Trinder, so impressed Hylton that he was signed by him. The 1920s also saw the winter garden converted into the Palace of Fun amusement centre. The Palace Pier Theatre often staged West End plays, whilst the bandstand still played host to military bands. Open-air dancing was held at the end of the pier, and there were firework displays every Saturday evening during the season.

In 1929 the theatre's heating and ventilations systems were upgraded, at a cost of £300, and the 1930s saw further improvements being carried out to the pier. In 1930 the entrance was set back 40ft due to seafront widening, and it was remodelled with a canopy and clock tower designed by Clayton & Black and erected by W.G. Beaumont & Son of London, at a cost of £1,909. Four years later, the pier's managing director, Oliver Dalton, re-sited the Chain Pier's tollhouses on the Palace Pier. In 1935 a dodgem car track was built as an extension over the landing stage, and two years later the East Pavilion sun terrace was added. A big wheel was also erected at the end of the pier. In 1939 concerts by Colin Mann on the Lafleur organ were a new feature.

Sadly, in October 1939 Oliver Dalton committed suicide. He was succeeded by Winifred, his wife, who remained at the post until her retirement in 1973. On 23 May 1940 a section of the pier was blown out as a defence measure – the order had been given whilst the audience queued for a performance of *The First Mrs Fraser*. The pier was also heavily mined, and a Hotchkiss heavy machine gun placed on the roof of the theatre.

The pier was reopened on 6 June 1946, and soon regained its popularity. In 1947 it was featured in the film version of Graham Greene's novel *Brighton Rock*, and the theatre featured in Richard Attenborough's film, *Oh! What a Lovely War*, in 1968, which was shot mainly on the West Pier. Visiting repertory companies and music hall revival shows became the main stay of the theatre, and summer seasons were held by established stars such as Dick Emery. Fishing competitions had been a feature of the pier since it was opened, and remained popular until angling from the pier was banned in 1984. During the 1970s the pier had over forty retail outlets selling all types of confectionary and souvenirs.

On 19 October 1973 the pier was badly damaged when a barge in use by workmen demolishing the landing stage broke loose from its moorings at the end of the pier during a gale and was driven repeatedly against the pier head girders. As the supports snapped, part of the decking collapsed into the sea, taking with it a phone box, first-aid post and the helter-skelter (which had been featured two years earlier in the film *Carry On At Your Convenience*). The north-west corner of the Pier Theatre was left hanging precariously over the sea, declared unsafe and closed. The cost of the damage was estimated at £1 million (for which the pier was not insured), and although the pier was eventually repaired, the theatre was never reopened. In 1986 it was removed to make way for an amusement dome, which itself was later cleared away to make room for a funfair.

In March 1984 the pier was sold by Brighton Marine Palace & Pier Co. to the amusement group, the Noble Organisation, for £1.5 million. They soon introduced free admission and deck chairs on the pier, and banned angling. A new façade was added to the front of the Palace of Fun and, in 1994–95, the pier head was enlarged by 92ft, over the site of the demolished landing stage, to enlarge the deck area for the funfair. Twelve new rides were controversially introduced in 1999, and following a public enquiry two 40ft roller coasters were erected in 2003.

The Nobles' controversially renamed the pier 'Brighton Pier' in 2000, notwithstanding the fact that Brighton's other pier, the West Pier, was still standing, albeit derelict. The company's opposition to the restoration of the West Pier using Heritage Lottery funding (which they regarded as unfair, that a future commercial rival had been awarded the subsidy) also rankled with some.

The pier suffered a fire on 4 February 2003 which saw the ghost train, log flume and mini bumpers destroyed, and other rides such as the helter-skelter damaged. The fire started at 7.20 p.m. and flames were seen shooting 30ft into the air. Fortunately, the wind was blowing the flames out to sea, and the fire had been contained by 9 p.m. Aside from some burnt-out areas of decking, there was no other structural damage to the pier and it was quickly repaired. In 2007 the ghost train was replaced by the House of Horror.

The pier is one of the most visited free attractions in the United Kingdom and, since 1996, has even been licensed for weddings. Attractions include its own resident DJ, Horatio, and Victorian bars, fish and chip restaurants and other food outlets, amusements and the funfair. The structure is well maintained and, with its delicate ironwork and attractive buildings such as the Chain Pier kiosks, still provides an elegant outline. Unsurprisingly, the amusement park at the end of the pier, and the amusements in the former winter garden make the pier very popular with younger people.

7

BRIGHTON WEST PIER

The Brighton West Pier Co. was formed in 1863 'to erect a handsome, commodious, and substantial iron promenade pier in the centre of that portion of the Esplanade which, at all seasons of the year, is the most thronged by residents and visitors'. The original capital of £20,000 (£2,000 × £10 shares) was soon increased, later in 1863, to £25,000 (2,500 × £10). The design of Eugenius Birch was chosen for a pier 1,115ft in length with a 140ft × 310ft pier head. Robert Laidlaw of Glasgow was the chosen contractor, and work began in March 1864 at an estimated cost of £21,840. The work was expected to take around twelve months, but in fact took two and a half years – the delay was said to be due to problems with shipping the iron components from the foundry in Glasgow. Rising costs and under-funding soon led to financial difficulties for the Brighton West Pier Co., and in August 1866 it was re-incorporated with an increased capital from £25,000 to £35,000. On 6 October 1866 the pier was officially opened by Henry Moor, chairman of the Brighton West Pier Co., in the presence of the Mayor of Brighton. A twenty-one-gun salute was fired by the coastguard, and then the 140 dignitaries were treated to a grand dinner in the banqueting room of the Royal Pavilion. The day's celebrations were concluded with a display of fireworks.

The entrance to the new pier featured two large square tollhouses, which were not appreciated by the residents of Regency Square opposite, who claimed that they spoiled their sea view. Admission onto the pier was originally 2d (1d on Sundays), but this was later raised to 6d to keep the pier select. Six ornamental kiosks in vague oriental style were placed on the pier and were used as gift shops, refreshment rooms and lady's and gentlemen's rooms. They each had a small spiral staircase leading to a minaret roof from which a panoramic view of the seafront could be obtained. Bench seating was provided around the edge of the pier, and the attractive cast-iron gas lamps, keeping with the oriental theme, had a serpent design. Ornamental weather screens were placed on three sides of the pier head, which had a small raised platform for band concerts. In 1893 the weather screens were relocated along the neck of the pier to make way for the pavilion.

The pier immediately proved popular, but in the summer of 1867 the public's confidence in the structure was shaken when it was felt to vibrate a number of times, causing customers to rush off the pier in panic, thinking it would collapse. It was thought that a steamer using the pier as a 'brake' might have been the cause. However, there was real concern over the stability of the pier, and it was strengthened.

In 1875–77 the centre of the pier was widened and a small covered bandstand was added. This was joined, in 1888, by a covered orchestra stand on the pier head, which had moveable awnings for sheltering the audience in bad weather. A further attraction was the organised displays and competitions held by the Brighton Swimming Club, and there were aquatic sports and entertainments. Miss Louie Webb performed underwater feats in a glass tank placed in one of the kiosks, and there was also a flea circus.

Brighton West Pier pictured on a regatta day soon after it was opened in 1866.

A fine view of the West Pier as originally built, taken from Regency Square. Clearly seen are the shore-end upper level, the six kiosks and the wind screen at the end of the pier.

Taken from a stereograph card, we see the primitive early band platform on the head of the West Pier.

In 1875–77 the centre of the West Pier was widened and a proper bandstand was erected.

During the storm of 4 December 1896, timbers from the wrecked Chain Pier wreaked vengeance on their old rival, the West Pier, by smashing through its supporting piles.

A fine view of the sea-end of the West Pier (c. 1906) showing the pavilion, erected in 1893, which was converted into a theatre in 1903. Also seen are the landing stage and bathing station.

A postcard of the interior of the West Pier Theatre, posted in 1906.

A crowd on the West Pier watch the 1906 Brighton Regatta.

The West Pier seen from the shore in 1906 with a steamer docked at the landing stage.

A rare photograph of the West Pier kiosk situated underneath the entrance to the pier. Ices were available at 2*d*, 4*d* and 6*d*; ham sandwiches, cheese rolls and hard boiled eggs at 2*d* each.

A feature of the Edwardian years on the West Pier was James Doughty and his performing dogs. Describing himself as the 'oldest living clown and actor', Doughty is seen here in his ninetieth year, in 1907.

The Lynton Troupe of trick cyclists on the West Pier, c.1907.

Prof. Powsey's Bicycle Dive from West Pier, Brighton.

Performing divers were a feature of the West Pier up until the Second World War. Here we see Professor Powsey's bicycle dive from the pier, *c.* 1905.

The West Pier Concert Hall under construction in 1915.

WEST PIER BRIGHTON.

Above: The West Pier in its heyday during the 1920s, following the addition of the Concert Hall.

Right: The programme of music from Paul Belinfante and his Quintette, in the Concert Hall on 10 November 1924.

WEST PIER CONCERT HALL BRIGHTON

The golden-haired, rather eccentric and temperamental H. Lyell Taylor, conductor of the Brighton West Pier Orchestra between 1918 and 1921.

Zoë Brigden was a local Brighton girl who dived off the West Pier from 1916 to 1925. Her most popular dive was the 'wooden soldier'.

A postcard showing the *Graf Zeppelin* flying over the West Pier. In 1936 the airship was seen in a number of resorts along the South Coast.

The West Pier in the 1930s. The race track at the shore end was added in 1927.

In 1949 the Concert Hall was reopened as a café with musical accompaniment.

In 1968 the West Pier was used extensively by Richard Attenborough for his film version of *Oh! What a Lovely War*.

The sea-end of the West Pier was closed off as unsafe in 1970. This view of the area was taken on 25 February 1981. The rest of the pier was closed in 1975.

The West Pier Theatre in 1982 when it was still in comparatively good condition.

By 1997, when guided tours of the pier were taking place, twenty-two years of dereliction were all too obvious to see.

Taken on 29 December 2002, crowds on the beach are seen gathering wreckage from the West Pier after part of it collapsed the previous day.

Hopes of restoring the West Pier were struck a cruel blow when the theatre was destroyed in a suspected arson attack on 28 March 2003.

A further arson attack burnt down the West Pier's Concert Hall on 12–13 May 2003.

Above: The gaunt, skeletal remains of the West Pier in April 2005.

Left: The future of the West Pier appears to lie in the success of the Brighton i360 Tower to be built at the entrance to the pier.

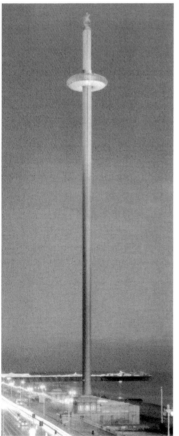

The pier initially proved to be financially successful, with dividends of up to twelve per cent being paid to shareholders. However, from the mid-1880s, the dividends declined as visitor numbers fell. This led to a reformed Brighton West Pier Co. being incorporated, in 1890, with a share capital of £100,000 to invest in new attractions. Eugenius Birch's nephew, R.W. Peregrine Birch, was appointed civil engineer, and he put forward proposals to alter and widen the pier and build a new landing jetty. The pier head was widened by 8oft, and on 19 October 1893 a 125 × 100ft iron-framed pavilion, complete with a colonnade of shops, was opened, although Birch's original proposal for the building to have a large central dome was refused by Brighton Corporation. Around 1,200 people could be seated in the pavilion, which had its own orchestra and choral union. Amongst the other fares offered was English light opera, operetta companies, musical sketches, mimics, female impersonators and ventriloquists. The pavilion proved a success in bringing more people onto the pier and dividends were back up to nine per cent for shareholders in 1898–99.

A landing stage of mild steel was erected around the three sides of the pier head between 1893 and 1896, and it was enlarged in 1901. The enclosed area of water between the pier head and the landing stage was used as a bathing station and a performing area by the pier's famous array of divers. These included Professors Powsey, Reddish and Cyril. The latter, who's real name was Albert Huggins Heppell, was fatally injured performing his 'sensational bicycle dive' on 27 May 1912 after suffering a side-slip and being thrown heavily onto the deck of the pier where he fractured his skull. R.L. Ede and Walter Tong were others who dived off the pier during the Edwardian period. Tong, who was sometimes assisted by May Victoria, was described as a 'high and fancy diver and ornamental swimmer' and a 'professional diver and life-saving champion'. He was said to specialise in the Moleberg and 50ft dives.

A fairly intensive service of steamer trips was operated from the pier's new landing stage, particularly by P&A Campbell. Their *Brighton Queen* operated from both the West and Palace Piers to Hastings, Eastbourne, Bournemouth, the Isle of Wight and France. Other visitors included *Brighton Belle*, *Glen Rosa*, *Glen Gower*, *Ravenswood*, *Albion*, *Devonia* and *Waverley*.

A feature of the pier between 1899 and 1903 were the Mohawk Minstrels, a large troupe that incorporated up to forty members. Their attire usually consisted of blue coats with red and white buttons, white shirts with large collars, huge bow ties and red and white striped trousers. The group played banjos and concertinas whilst sat in a semi-circle with their compere, Mr Interlocutor, in the centre.

In 1903 the pavilion was converted into a 1,076-seat theatre, which was utilised all year round. Touring companies were often a feature, receiving forty-five to seventy five per cent of the receipts. Concert parties which appeared there included Harry Gold's Brighton Cadets and Fred Austen's Merrie Middies & Maids. Sacred concerts were held on Sundays, until the First World War.

Out on the pier deck, Professor James Doughty, who at ninety years of age was the self-proclaimed 'World's Oldest Living Clown', entertained with his performing troupe of dogs who wore hats and coats, jumped through hoops, balanced on large balls and danced on their hind legs. Doughty's wife Alice was only twenty-five years old, and when he passed away at the age of ninety-four, in 1911, it was said he had a smile on his face! The Lynton Troupe of trick cyclists was another popular attraction on the pier deck.

Following the addition of a winter garden on the Palace Pier in 1911, a second concert venue was also planned for the West Pier. Work began in 1914 with the removal of the bandstand and the widening of the centre of the pier by 14ft. The Concert Hall, as it was termed, was a low eight-sided oval building erected around iron arches designed by Clayton & Black and Noel Ridley. On 20 April 1916 it was officially opened with a concert by the King's Royal Rifles Silver Band. Seating 1,300 people, and with a bandstand in its centre, the hall was home to the West Pier Orchestra, conducted between 1918 and 1921 by the golden-haired, rather eccentric, H. Lyell Taylor, who was fond of telling the audience off if they in any way interrupted or made a noise during his performance.

Following the end of the First World War, visitor numbers to the pier soared, and in 1920 there were 2,074,000 paying customers. The pier was used for the sporting and bathing costume pageants of the Brighton Carnival, and divers were still a popular attraction, particularly two lady performers, Zoë Brigden and Gladys Powsey. Zoë was from a local Brighton family and was the town's amateur swimming champion for a decade, until turning professional in 1913. She turned to diving following the end of her swimming career through a shoulder injury, and first performed on the West Pier in 1916, in partnership with Walter Tong. In 1918 she went solo and continued to perform on the pier until 1925, regaling people through a megaphone to come and see her show. One of her most popular dives was the 'wooden soldier', where, with her arms held tightly to her side, she plunged head first into the sea. Gladys Powsey was the daughter of Professor Powsey, who had dived off the pier before the First World War. With her swimming costume emblazoned with the sponsorship of Bovril on her chest, Gladys included in her act an imitation of a seal swimming, diving and calling out. She later moved to Herne Bay, where she dived off the pier there. In the 1930s Ray Brooks, who styled himself as 'Aquamaniac', dived off the West Pier.

However, by 1930, visitor numbers were on the decline, and in 1939 they were less than half of that recorded in the 1920 boom. The pier was being hit not only by the more centrally placed Palace Pier but also by cinemas, dance halls, open-air swimming pools and other attractions in the town. The Concert Hall began catering for the mass market, staging less 'stuffy' kinds of music by dance bands, light orchestras, military bands and concert parties such as the Follies, West Pier Vanities and Cyril's Celebrity Concerts. The Pier Orchestra was greatly reduced in size and renamed the Pier Bijou Orchestra. The West Pier Theatre was closed during part of the winter and began to be used for other purposes such as lectures. The play *The Ghost Train*, by Arnold Ridley, returned to the theatre on a number of occasions, as did *No No Nanette*, *Pygmalion*, *Charlie's Aunt* and *The Merry Widow*.

During the 1920s amusements began to have a bigger role on the pier. In 1927 a miniature race track was added at the shore-end, and an amusement arcade was opened in the 1930s. Income was also derived from letting kiosks and advertising spaces. From the 1920s to 1940: Harold Stoner paid 1s per week to work as a sand artist just to the east of the pier; J.E. Hackman paid 10s a week for using his jockey scales on the pier; Madame Stellar paid £84 per annum for her palmistry studio; and there were leases for the speedboat, auto golf, silhouette artist, shooting gallery and amusement arcade. The Punch and Judy man on the pier was Tom Kemp, who was said to have kept a live crocodile in his bath! The pier company received from the concession holders either a fixed monthly rent or a percentage of the gross receipts.

Improvements to the pier included a new raised entrance at the shore-end, in 1932, and a bathing station and sun terrace on the eastern side of the landing stage, in 1935. Tickets cost 6*d*, including admission to the pier. However, with the coming of war four years later, the pier had a 60ft hole blown in it beyond the Concert Hall, as a defence measure, and the pier head and landing stages were mined. The first section of the pier up to the Concert Hall was reopened on 18 April 1946, and, once the gap in the structure was restored, the remainder followed suit in 1948. To carry out the repairs pending their war damage settlement claim, the Brighton West Pier Co. borrowed £90,000. The war damage claim was settled in 1948.

Following the restoration of the pier, the management decided to put both the Concert Hall and theatre to new uses. The Concert hall was converted into a café with musical accompaniment, while the theatre was subdivided into the Ocean Restaurant upstairs and the Laughter Land Games Pavilion downstairs. In 1947 P&A Campbell paid £300 for the use of the landing stage by one of its vessels, either the *Glen Gower* or *Empress Queen*, although regular sailings ceased in 1956. The diving tradition was continued by Professor Javlin in the 1950s, and The Great Omani in the 1960s, who performed his 'Death Dive', based on an original Houdini act, by diving into the sea all chained up. He also carried out his 'Fire Dive', where he jumped from the pier into a ring of fire in the sea.

Other attractions included the Richold Collection of matchstick models; the star feature being a model of Milan Cathedral that used 8,000 matchsticks and took five years to build, between 1893 and 1898. Bathing continued to be enjoyed at the pier head, and there were regular speedboat trips from the landing stages. Amongst the concessions on the pier were jockey scales, shooting gallery, portrait sketches, fancy goods, shellfish and fruit stalls. The car race track was still functioning inside the pier entrance, and other amusements included the dodgems, ghost train, a crazy maze and a roundabout.

During the early 1950s the Brighton West Pier Co. acquired a provisional order in an attempt to raise additional capital by borrowing money, levying increased rates and charges, and raising the pier toll from 4*d* to 6*d*. A new pavilion was planned at the shore-end of the pier, although this was never built. In 1952 the theatre suffered damage by fire.

Although numbers using the pier continued to fall, it remained reasonably successful, and in the Concert Hall Alan Gale's *Old Tyme Music Hall*, first performed on 25 June 1964, proved very popular. Also, as already mentioned, in 1968 Richard Attenborough used the pier extensively for his film version of *Oh! What a Lovely War*. However, in 1965 AVP Industries, who owned the Metropole and Bedford hotels in Brighton, gained a ninety-seven per cent holding in the pier company, and announced plans to turn it into a luxury holiday and conference centre. This failed to happen, and in September 1969 AVP announced their intention to demolish the theatre and remove the two entrance pavilions. However, Brighton council refused to give listed building consent for what was a Grade II structure. AVP responded by declaring the pier head housing the theatre to be unsafe, and closed it at the end of the 1970 summer season. They claimed £655,500 was needed to restore the sea-end of the pier, and a further £108,000 to repair the remainder. During the year ending 28 February 1970, AVP declared a net loss on the pier's operation of £6,173, which, with accumulated deficiencies from previous years, amounted to £29,166. The pier toll in 1970 was 9*d* for adults and 6*d* for children. In July 1971 the Department of the Environment gave listed building consent for the demolition of the pier head, subject to the approval of Brighton council, and the remainder of the pier remaining intact. Nothing was done, however, and AVP and the council continued to debate what should be done.

By January 1974 a survey had revealed that £1 million was needed to restore the pier. At the end of the year the decision by the policy and resources committee of the council to no longer oppose the demolition of the pier head led to the foundation, by John Lloyd, of the 'We Want the West Pier' campaign. Their strong campaigning sparked intense local media interest, which led to a full meeting of the council overturning the decision of their policy and resources committee. However, the council was powerless to prevent the closure of the whole pier by AVP on 30 September 1975.

Early in 1977 the Brighton West Pier Co. was wound up following the failure of AVP to provide funding, and in the following year the pier passed into the hands of Crown Estate Commissioners. 1978 also saw the formation of the Brighton West Pier Society (later the Brighton West Pier Trust) and a parliamentary order in 1979 allowed the society to operate and maintain the pier following purchase. In 1982 the pier gained Grade I listed status, and two years later the Brighton West Pier Trust took over the pier from the commissioners for a nominal sum of £100. By 1985 the cost of restoring the pier had risen to £8.5 million, and in the following two years emergency restoration works were carried out under the seaward end and Concert Hall.

Various proposals to restore the pier came and went, although a small part of the shore-end was reopened in 1987 at a cost of £230,000. Unfortunately, damage sustained in the Great Storm of 16 October that year led to a 110ft section of the pier between the restored entrance and the Concert Hall having to be dismantled, leaving the Concert Hall and theatre cut off from the shore.

However the Brighton West Pier Trust battled on and salvation for the pier appeared to come with the granting, in 1996, of £14.1 million of National Lottery money for its restoration. £1 million was handed over immediately, for essential repairs, and a temporary walkway was laid along the pier enabling guided tours to take place. Costing £15 per head, they proved to be very popular. The total cost of full restoration of the pier was estimated at £30 million (£15 million up to deck level and £15 million on decking and buildings), of which the Lottery money would pay for half. Private funding was sought for the remainder. A deal was reached with the developers St Modwen, and as part of the restoration new buildings were planned adjoining the entrance to the pier. The planned reopening of the pier was set for New Years Eve 1999.

However, by 2002 the restoration and redevelopment of the pier had not commenced, bogged down with objections from the 'Save our Seafront' pressure group and the Regency Society to the seafront buildings proposed by St Modwen. Furthermore, the owners of the Palace Pier, the Noble Organisation, were opposing the restoration, claiming it was unfair to use public money to restore a rival attraction. No Lottery money had been handed over, and the public tours had ceased due to the ever-deteriorating condition of the pier. On 29 December 2002 and 20 January 2003 storms led to the partial collapse of the Concert Hall and the 1996 walkway. Worse was to follow when suspected arson attacks led to the destruction of the theatre on 28 March 2003 and the Concert Hall on 12-13 May 2003.

Brighton and Hove city council still maintained its support for the pier's restoration, having voted in February 2003 in favour of the controversial plans for the new seafront buildings. The National Lottery Heritage Fund felt differently, however, and on 28 January 2004 announced that the £19 million earmarked for the refurbishment was to be withdrawn, claiming there was virtually nothing left now to restore. Their decision effectively ended hopes that the pier could be saved. English Heritage also withdrew their support following the collapse of the skeletal remains of the Concert Hall during a storm on 23 June 2004. In the following August St Modwen announced that they were pulling out of their deal with the Brighton West Pier Trust.

All that effectively remained standing of the pier was the tiny section of restored entrance and the skeletal remains of the theatre upon the pier head. The starlings continued to swoop around the bare iron supports of a once grand Victorian building that, even in its naked state, continued to exert a fascination of its own. With its hopes of restoring the pier effectively ended for the present, the Brighton West Pier Trust looked at other ways of keeping interest in the pier's eventual restoration. In 2006 they announced plans to build the Brighton i360 viewing tower at the entrance to the pier, which were supported by Brighton & Hove city council. The tower will be 500ft high, and up to one hundred visitors will be carried to the top in a 4m-wide aerodynamically designed pod. From the observation deck outstanding views of the English Channel and South Downs will be gained. Around twenty per cent of the tower's energy needs will be generated by silent wind turbines on top of the spire. It is hoped that the tower will attract around 600,000 visitors a year, and its success will provide funds for the eventual restoration of the pier, albeit in a different style to its predecessor.

8

WORTHING PIER

Worthing was just a small fishing settlement until the end of the eighteenth century when it began to gain some popularity as a sea bathing centre visited by people from nearby Brighton. Peter Wycke of London is said to have been the first visitor who came to Worthing for his health in 1759. Bathing machines were introduced in 1789, and by 1813 there were sixty available for hire. The Steyne was erected on the seafront, and the infant resort received the boost of a Royal visit when, on the advice of George III's physicians, Princess Amelia stayed there for four months in 1798. In 1803 Worthing was granted town status, and four years later a second Royal visit occurred when Princess Charlotte, the daughter of the Prince Regent, stayed at Warwick House. John Evans published his guide *Picture of Worthing* in 1805, and in 1807 a theatre was added to Worthing's attractions, which also included baths, circulating libraries and an assembly room. Between 1819 and 1821 Worthing's first primitive promenade, consisting of rolled sea gravel and sand, was laid from West Buildings to Warwick Road. During the 1820s the grand terraces of York Terrace, Liverpool Terrace and Park Crescent were developed, yet the town experienced a slump during this period. However, the arrival of the railway in 1845, and the foundation of a local board seven years later, re-invigorated growth, and Worthing took on the guise of an unpretentious resort for the middle classes.

Worthing was one of the earliest resorts to advocate building a pier, and in 1860 the Worthing Pier Co. was formed with a capital of £5,500 in £1 shares, which was soon increased to £7,500. The pier design of Sir Robert Rawlinson was selected, which envisaged a pier of 960ft in length, standing 10ft above high water. Joseph Cliff of Bradford was the chosen contractor, and he accepted 500 shares in the company. Rawlinson had 400, although by 1868 he held 600. The first pile of the pier was driven in by Cliff on 4 July 1861, in wet and stormy conditions, and the pier was officially opened on 12 April 1862.

The finished structure was a basic promenade pier with little ornamentation. There were no buildings on the pier, although bands were engaged to play on the open pier head. By 17 February 1863 all the shares in the company had been taken up.

The first Pier Master, Mr Parkes, was replaced in 1869 by Henry Hayden. Mr Hayden was born in Newport, Isle of Wight, in 1810, the son of a ship's master who owned his own trading brigantine. After completing twelve years service in the Royal Navy he married Ann Gerard, the daughter of a baker from Kemp Town, Brighton, who was of Huguenot stock. Hayden entered the coastguard service and, after being employed at various stations along the Sussex coast during a period of seventeen years, came to the Worthing coastguard station in 1857, where he remained until taking up the post of Pier Master in 1869.

In 1874 nine musicians were engaged to play on the pier for three hours a day, weather permitting, whilst in 1881 a German Rhine Band were contracted to play on the pier daily.

Improvements were carried out to the pier in 1884 when two kiosks were added at the landward end by Alfred Crouch, which housed a tollhouse and a fancy repository that sold souvenirs. Three years later the directors of the Worthing Pier Co. instructed the eminent civil engineer, Mr James Mansergh of Hampstead, to prepare a large-scale redevelopment of the pier. The scheme was approved by the shareholders in 1888 and the company was re-formed with an authorised capital of £20,000 in 40,000 shares of 10s each. The projected alterations and improvements included the doubling of the width of the pier to 30ft for a distance of 750ft and replacing a small structure on the pier head with a pavilion. The contract for the ironwork and widening of the structure amounted to £5,860, and was awarded to James Hobbs and John Wright. The construction of the pavilion and landing stage brought the total outlay up to £12,000.

The enlarged pier was officially opened by Viscount Hampden, Lord Lieutenant of Sussex, on 1 July 1889. Following the opening ceremony a grand public luncheon was held at the New Assembly Rooms in Bath Place.

Overseeing the revamped pier was Pier Master Tom Belton, who had succeeded Henry Hayden in 1887. Belton was a local man, born in Marine Place in 1835 to a local fisherman. In 1850 he survived a disaster at sea which saw eleven of his fellow fishermen drowned. Married at the age of thirty, he quit the fishing industry and signed on as a merchant seaman, sailing between British ports at first, and then on longer voyages all over the world. He eventually returned to Worthing, taking up fishing again for eleven years and crewing yachts before coming Pier Master. He was to remain in the post until 1908 when he was replaced by Irvine Bacon, the first non-seaman to be appointed.

The new pavilion was known as the Southern Pavilion and it was used for a variety of entertainments, including early moving pictures, when, on Monday 31 August 1896, Lieutenant Walter Cole presented a programme of Electric Animated Photographs. Pierrots and concert parties such as The Vaudeville Follies, Frank Major's Peers and Royale & Mervin's Middies & Maids were a regular feature (sometimes appearing on the pier's alfresco stage), and musical features included the Blue Hungarian Band, the Royal Navy Ladies Orchestra and the pier's own string orchestra. From 1910, Carl Adolf Seebold presented his Chamonix Orchestra, and roller-skating on the pier was extremely popular in 1910, bringing in almost as much income as the pavilion entertainments. Bathing was permitted from the pier head at a cost of 4d, and the pier structure was also used to help launch the lifeboat into the sea, which saw crowds rush onto the pier to witness the launches.

The pier's landing stage was not available at all states of the tide, but proposals to extend it were abandoned when it was learnt that even if the pier was twice as long, the water would only be 6ft deeper. The favourite vessel to call at the pier was the *Worthing Belle*. She was constructed as the *Diana Vernon* in 1885 by Messrs. Barclay, Curle & Co. of Glasgow for the North British Steam Packet Co. Powered by an 110hp single-cylinder steam engine; she was employed on their Holyloch and Gareloch service from Craigendoran. In 1901 the vessel was acquired by Mr J. Lee of Shoreham and renamed the *Worthing Belle*. She ran under the ownership of William Reid and the Brighton, Worthing & South Coast Steamboat Co. until 1913, when rival competitors forced her out. The *Worthing Belle* was then sent to Turkey and used as a ferry under the name of *Touzla* until she was scrapped in 1926.

Worthing Pier showing the original tollhouse and narrow walkway, c.1870.

In 1888–89 Worthing Pier was widened and a pavilion was added. This photograph was taken during the re-opening ceremony.

Worthing Pier during the Edwardian era featuring the entrance kiosks added in 1884 and the pavilion built in 1889. This postcard was sent on 31 May 1912, and the message reads: 'Dear Annie, what do you think of this lovely pier, best love "Yours".'

A close-up view of the attractive kiosks at the entrance to Worthing Pier with a couple of fine bath chairs in attendance, c.1905.

Local photographer Edwards & Sons provided this postcard of a children's carnival on Worthing Pier in 1908.

Frank Major's Peers photographed outside Worthing Pier Pavilion in 1912. They presented a variety show of music and comedy.

In the early hours of Easter Monday, 23 March 1913 almost the whole of the walkway of Worthing Pier was washed away during a fierce gale. This excellent postcard by Otto Brown shows the crowds gathered to view the pier at high tide. One of them was the sender of this postcard, who wrote: 'We have been crowded out with people to view this sight – motor cars by the score – the whole structure went with one dash by a big wave.'

The wreckage of Worthing Pier at low tide, on 23 March 1913, pictured on a postcard by Edwards & Son.

An unusual postcard showing the isolated pier head pavilion that survived the 1913 storm.

On 9 July 1913 the steamer *Worthing Belle* called at the stranded Worthing Pier head and deposited passengers who were rowed ashore in small boats.

Above: The isolated pier head was used as a backdrop to the Pashley Brothers' aircraft on the beach on 29 July 1913.

Right: The rebuilding of Worthing Pier following the 1913 storm. This postcard was sent on 23 March 1914, and the sender has commented: 'This is how far they have got with the pier.'

On completion of the rebuilding after the 1913 storm, Worthing Pier was officially reopened on 29 May 1914 by the Lord Mayor of London.

The new Worthing Pier of 1914 was widened at the shore-end to prevent another collapse happening again. A small stage can be seen on the pier in this postcard, and a concert is in progress.

On 25 June 1926 the Mayor of Worthing opened a new pavilion at the entrance to the pier. Designed by Adshead and Ramsey, the building cost £40,000 to construct.

An interior view of the new pavilion erected in 1926, which provided seating for 1,063 people.

On 10 September 1933 Worthing Pier suffered another calamity when the Southern Pavilion was destroyed by fire.

The rebuilt Southern Pavilion following the fire of 1933, designed in the art deco style of the time.

Worthing Pier with the wartime breach being repaired, *c.*1948.

The central amusement pavilion photographed in 1982. This was added to the pier in 1937, in an art deco style to match the Southern Pavilion.

The Southern Pavilion in 1982 when it housed the Pier Head Model Railway.

Worthing Pier photographed in 2006. The pier was a deserved winner of the National Piers Society's 'Pier of the Year' competition in 2007.

In 1909 a proposal was put forward to widen the shore-end of the pier and erect a concert hall and arcade. A pier order was granted the following year, and James Mansergh was to oversee the project, but, for the moment, it was not carried out.

All thoughts of alterations to the pier had to be put to one side following the dramatic events of Easter Monday 1913, as described here:

The gale that took place on the night of Easter Monday, March 22 1913, left behind a scene of unparalleled devastation. Though the increasing ferocity of the wind had deterred most people from going on the pier, a brave thirty souls had battled to the Southern Pavilion to listen to light music by the McWhirter Quintet. However halfway through the performance, the courage of the audience began to fail as the wind reached hurricane force and one by one they decided to reach the relative safety of dry land. As their audience deserted them, the musicians thought it wise to leave the pier, and clutching their instruments joined the many sightseers on land.

Around midnight the town was plunged into darkness as the street lights failed and the moon was hidden by clouds. Although they could barely see it, the sound of the waves thundering over the pier crashed out in a great noise, until, all of a sudden, the whole pier shook and then crashed into the sea, leaving the stranded Southern Pavilion out in the sea, which the locals affectionately named 'Easter Island'.

On the morning following the storm, huge crowds climbed amongst the twisted mass of pier wreckage on the beach. Small pieces were taken away as souvenirs and pennies from the wrecked amusement machines were eagerly snapped up.

The *Worthing Belle* continued to sail past the stranded pier head and ventured as close as possible so its passengers could have a good look. On Tuesday 9 July 1913 the vessel landed passengers onto the pier head, from where they were conveyed to the shore by local boatmen for a penny fare. During the regatta, some spectators paid the pier toll and were rowed out to the pier head where they spent the day watching the proceedings.

Work then began on rebuilding the pier, which was to be wider to give it extra stability. The pier was reopened with great ceremony by the Lord Mayor of London, Sir T. Vansittart Bowater Bart, on 29 May 1914. A public holiday was declared in the town and large crowds lined the procession route where the Lord Mayor travelled in his own horse-drawn carriage brought down specially from London. A luncheon was held at Warnes Hotel for the 176 dignitaries, who were entertained by the Royal Naval Ladies Orchestra.

The cost of rebuilding, however, crippled the Worthing Pier Co. and on 8 November 1919 an agreement was reached with Worthing Corporation to acquire the pier for £18,978. The sale was confirmed by the Worthing Pier Order of 1920. Six years later the council widened the pier entrance and erected the Pavilion Theatre at a cost of £40,000. Designed by Adshead & Ramsey, it was a single-storey stuccoed structure with curved metal-clad roofs designed to resemble the Southern Pavilion. The building housed the newly formed Worthing Municipal Orchestra, which became the only year-round municipal orchestra. The first musical director was E. Joseph Shadwick, from 1926–31. He was succeeded by John Heuval (1932–35), and then Herbert Lodge (1935–55).

Further misfortune hit the pier on 10 September 1933 when the Southern Pavilion was destroyed by fire. The blaze was first spotted by an off-duty fireman sunbathing with his family on the beach at around 3.30 p.m. He noticed wisps of smoke coming out from under the pavilion and ran to tell the fire brigade. Upon his arrival at the fire station he was surprised to learn that the brigade had not already been informed of the fire. By the time the firemen had arrived at the pier, the pavilion was well alight, the flames having been fanned by a strong southerly wind. Many holidaymakers, some still in their bathing suits, assisted the firemen by ripping up the pier's decking to stop the flames spreading along the structure. A car ran up and down the pier, taking away furniture from the blazing building. The smoke could be seen as far away as Beachy Head in the east and Selsey Bill in the west. It is thought the fire was started by a discarded cigarette end.

The replacement Southern Pavilion took advantage of the 1930s craze for sunbathing. Opened on 3 August 1935, the ground floor was equipped with a solarium fitted out with ultra-violet lamps and vita glass windows. A wide balcony ran around the first floor. The building cost £18,000, of which £13,717 was paid out of insurance. A central amusement pavilion in matching art deco 'jazz' style, sporting a stylish clock of the period, was added in 1937, along with a windshield down the centre of the pier.

During the Second World War the pier was breached as a defence measure, and the Pavilion Theatre was used as a recreation centre for troops, where they were entertained by films and concert parties. The pier had a lucky escape just after the end of the war when a mine that had drifted near to the structure was detonated by a bomb disposal unit.

The pier was reopened in April 1949 after the breach was repaired using adapted cast-iron water main pipes. The timber used to repair the landing stage used half of 1947's import of greenheart for the whole year!

The Pavilion Theatre hosted a number of popular variety shows during the 1950s and 1960s, including *Showtime, Show Box, Out of the Blue, Twinkle, Band Box, Sincerely Yours, Gay Parade, Evening Stars, Fol-de-Rols* and *Box of Tricks*. In March 1958 the building received a £9,000 facelift, and in July 1959 the Denton Lounge was built. An information kiosk was added in 1963, and a rather unsympathetic entrance canopy in 1970. Between March 1979 and April 1982 it was closed for extensive repairs and alterations, at a cost of just over £1 million. A Grade II listed structure, the Pavilion Theatre has seating for 850 patrons, and houses a restaurant and bar.

Still in the hands of the local authority, the pier is well maintained and was the deserving winner of the National Piers Society's 'Pier of the Year' competition in 2007. The Southern Pavilion and central amusement arcades are good examples of the art deco style of the 1930s.

9

LITTLEHAMPTON PIER

Littlehampton's little wooden pier at the harbour mouth dates back to 1735. This postcard dates from around 1910 when it was a popular place to sit and watch the shipping.

A look along Littlehampton's harbour pier in November 2000.

10

BOGNOR REGIS PIER

Bognor was just a tiny hamlet until a wealthy businessman and MP for Southwark named Richard Hotham stayed there in 1784 and decided to create a fashionable watering place which was to be called Hothamton. He acquired 1,600 acres of land and, in 1787, started work on his first residence, Bognor Lodge. Hotham spent most of his £100,000 fortune erecting forty elegant properties in the new resort, along with a plush new home called Chapel House (later Hotham Park House), before he passed away, on 13 March 1799, aged seventy-six. The debts left behind by Hotham were paid off by selling the development land: his great vision of a new Bath or Brighton having failed to materialise, although the new resort of Bognor (the Hothamton name not having stuck) did prove popular with the nobility. The Prince of Wales visited for several weeks in 1796 and his daughter, Princess Charlotte, spent her summers at the Dome, Hothamton Crescent, in 1808–11.

Development of the resort remained slow, however, as a number of schemes to expand it floundered with only a few buildings materialising. However, by 1829 the Norfolk Hotel had been erected and the Steyne and Waterloo Square were in the process of being developed. Princess Victoria visited her 'Dear Little Bognor' several times between 1821 and 1830, and helped retain its popularity with the nobility, particularly during nearby Goodwood's race week. The Marine Parade was constructed in the 1840s, and in 1846 a railway station was opened at Woodgate, before the line was extended to Bognor on 1 June 1864. By this time Bognor had established itself as a quiet and respectable seaside resort.

By the early 1860s a provision of a pier was being argued for Bognor. This led to the formation of the Bognor Promenade Pier Co. in 1863, and a pier was sanctioned by the Board of Trade in September of that year. Work began in 1864, and the pier was officially opened on 4 May 1865, in the company of the Band of the Royal Sussex Light Infantry. The pier had been constructed by J.E. Dowson, at a cost of £5,500, to the design of Joseph William Wilson. It was a simple structure consisting of a slender 13ft-wide neck, 1,000ft long, terminating in a 40ft-wide pier head. A report of the opening appeared in the *West Sussex Gazette* on 11 May 1865:

> The opening of the pier took place on Thursday last, and it is with much pleasure that we record a circumstance of such importance to the town of Bognor. An attractive pier is one of the best indicatives of future prosperity in such places, and is an unquestionable sign that the welfare of the town is being looked after by some of the inhabitants who have influence and means.
>
> Its beauty and healthfulness are positive claims which the town and neighbourhood are likely to reap from it. Bognor has for long been a very favourite watering place with a few of the select, and since it has received the advantages of a line of railway, connecting them with the

extensive and important system of the London, Brighton and South Coast Railway Co., it has gradually but quite perceptively increased in importance.

It is now about eighteen months since a few gentlemen started the movement to obtain this much desired object and it has throughout been a matter of surprise to the originators that a greater amount of interest was not manifested, and that so little support should have been rendered them in their undertaking.

In April of last year the first pile was driven, and about five weeks since the erection was completed. It is exceedingly pretty and has a chaste and light appearance, and from the most reliable information we could glean, we believe it to be a substantial piece of work.

The pier was constructed purely as a marine promenade, and entrance for strollers cost 1d (plus 4d for bath chairs). The first Pier Master was John Smith, who held the post for twenty-five years. He was a much-respected figure in the town, and on his death all the blinds in the town were drawn and the pier was closed for the funeral. Mr Smith was succeeded in the post by his brother William.

The Bognor Pier Co. struggled to make the pier pay, and in 1867 slid deeper into financial trouble after the sloop *Nancy* slipped her moorings and hit the pier. Eventually, on 6 December 1876, the local board purchased it for £1,200, and in 1880 added a bandstand. However, they also failed to make the pier pay, and there were complaints about the poor condition of the structure. In January 1890 the board informed the Board of Trade that they were hoping to find a purchaser and dispose of the pier, as they could not make a go of it purely as a promenade. They termed the pier the 'white elephant' that never covered its expenses: the average annual income of late had been £174, while expenditure was £285. They also still owed £450 on the £1,500 borrowed in 1876 to purchase the pier. The pier was closed for the winter and repairs were carried out to defective crossbeams and decking. On 12 August 1891 a provisional agreement was reached with the engineer Frank Kirk to improve the pier and then take it over. The completed deal was to take place by July 1892, but this failed to take place after Kirk objected to the terms of his takeover, which limited the tolls he could charge, and thus the profits he could make, to ten per cent. On 5 March 1894 the Bognor Pier & Pavilion Co. Ltd was incorporated with a capital of £5,500 (5,500 × £1 shares) to acquire and improve the pier, but they were wound up two years later having achieved nothing.

In 1895 pier bandmaster Mr Howes agreed to supply a band, at a cost of twenty-three guineas per week, twice daily. However, on 9 July 1900, the council erected a permanent pavilion on the pier head, and three years later added a small landing stage. Nevertheless, steamers found it difficult to use because of the tides, even after reconstruction in 1935. A diving platform was placed on the landing stage, and divers such as Professor Ralph Owen, Neptune and Nereid (1913), Madge Goodall (1920s), Professor Davenport (1926) Angus Kinslea (1930) and Ambrose Smith (1950s) were a great attraction.

The pier made small surpluses in 1904 and 1906, but a loss of £209 in 1905 due to the cost of repairs. In 1907, the last full year of council ownership, a small surplus of £83 was gained, with income derived from tolls (£324), rent on the pavilion (£10) and automatic machines (£35). Expenditure included the wages of the pier keeper and manager (£72), repairs and other wages (£85) and bands and other entertainments (£46).

The opening day of Bognor Pier on 4 May 1865, as featured in the *Illustrated London News*.

A busy scene outside the original entrance to Bognor Pier during the Edwardian period. This was to be swept away in 1909 when work began to greatly improve the pier. (Marlinova Collection)

A photograph of Bognor Pier in the 1880s, looking from the pier head to the shore. The pier at this time was just a bare promenade with a couple of shelters and some bench seating.

A postcard of Bognor Pier in 1902 showing the pier head pavilion added two years earlier.

A group of people watch a paddle steamer off Bognor Pier, c. 1905. The pier's small landing stage had been added in 1903.

Parade & Pier, Bognor.

The shore-end of Bognor Pier pictured on a postcard sent on 25 November 1907. A simple tollhouse graced the entrance to the pier.

In 1908 Bognor Pier passed from the local council to Alfred Carter and Michael Shanley. They set about improving the structure, and in 1910–12 added a theatre and cinema at the shore-end. This photograph shows the work in progress.

The new picture hall and theatre on Bognor Pier in 1912. The work to widen the pier and erect the buildings cost a total of £30,000.

The interior of Bognor Pier Theatre in 1912, which could hold a capacity of 1,400 people. The building was designed by G.C. Smith and erected by W.H. Archer & Son.

People rush along Bognor Pier to catch the waiting steamer, c. 1912. During the winter the pier head pavilion was used for roller-skating.

Above: A postcard issued by King & Wilson, 8 Pier Arcade of Bognor Pier, *c.* 1913. Another of the pier's attractions, the rifle range, can be seen advertised.

Right: Madge Goodall performing her sack dive off Bognor Pier in the 1920s.

The *Graf Zeppelin* flying over Bognor Regis Pier in 1936.

A view of Bognor Regis Pier taken from the diving stage in 1936.

The illuminated deck and pier head pavilion of Bognor Regis Pier in 1936. The diving stage can be seen on the landing stage.

Bognor Regis Pier from the air in 1936 showing the landing stage added in that year.

The Second World War breach in Bognor Regis Pier.

A postcard sent on 1 September 1953 featuring the miniature railway laid on Bognor Regis Pier by Charlie Kunz. The locomotive was a replica LNER 462 Pacific.

The broken neck of Bognor Regis Pier in 1982. The gap in the pier had been caused by storm damage during the winter of 1964–65, when the old pavilion fell into the sea.

A photograph taken on 24 October 1999 showing the destruction by storm of the section of Bognor Regis Pier that had lain derelict for a number of years.

From 1979 to 2007 the Bognor Birdman Rally was held on the pier, attracting competitors from around the world. However, a further shortening of the pier in March 2008 meant the event was transferred to Worthing (although it was subsequently cancelled due to bad weather).

Bognor Regis Pier photographed in March 2004. The three end spans were removed in March 2008.

Finally, on 2 October 1908, the pier was sold, for 10s 6d, to Messrs. Alfred Carter and Michael Shanley, who had an established catering and deck chair businesses. Shanley was known as the 'Deck Chair King' because he supplied them to so many resorts, as well as London parks and for special occasions. Carter and Shanley formed a new Bognor Pier Co., which was incorporated on 10 April 1908 with a capital of £10,000 in £1 shares, and immediately spent £2,565 on repairs to the pier. The work was used as a basis for a film by Cecil Hepworth called *Painting the Pier*. On their first night of ownership films were shown in the pier pavilion, but between 31 October 1908 and 3 April 1909 the pier was closed for repair work. This was just a prelude to a major reconstruction of the shore-end of the pier in 1910–12, at a cost of £30,000. The pier was widened to 80ft, for 300ft of its length, by Messrs. Weston & Burnett of Southampton, and a theatre and cinema complex with twelve shops was erected by W.H. Archer & Son of Gravesend, to the designs of G.C. Smith.

While the work was taking place, roller-skating was the main attraction on the pier, and during the summer a rink was laid on the pier deck, which attracted many complaints about the noise. The pier head pavilion showed Shanley's High Class Animated Pictures.

The 528-seat cinema was opened in 1910, and the theatre the following year. This could seat 1,180, and featured a roof garden where the Bijou Orchestra performed daily and where afternoon tea was served. Mr Flude was appointed entertainments manager (a position he was to hold for fifty-five years, until 1964), and London theatre companies were engaged to perform nightly. There was also a concert deck at the back of the theatre for Pierrot troupes and concert parties, and in 1913 a children's circus in the old pavilion and a rifle range were added.

The improvements carried out by the new owners immediately made the structure profitable. Between 1909 and 1914 there were net profits of £192 (in 1909), £239 (in 1910), £245 (in 1911), £140 (in 1912), £625 (in 1913) and £272 (in 1914).

During the First World War army personnel were billeted on the pier, although it remained open to the public and proved to be very popular. Good net profits were made in particular in 1917 (£1,149) and 1918 (£1,378). In 1918 income was gained from the cinema (£4,958), theatre (£1,547), automatic machines (£236), rifle range (£97), pier tolls (£568), rent of shops (£145), confectionery and refreshments (£634) and bar sales (£225).

During the 1920s the Boquets concert party became a feature of the pier during the summer season, and changed their programme a remarkable five times per week. In 1929 Bognor was put firmly on the map when King George V stayed at Craigwell House while recuperating from a lung infection. The local council took advantage of the visit by asking for Royal recognition, and they were granted the suffix 'Regis' on 1 June 1929.

Bognor became increasingly popular with day trippers during the 1930s, and the shore-end buildings of the pier were further enlarged in 1934–36. On 10 June 1936 a new landing stage, designed by Oswald Bridges and built by Jackman & Son, was opened, measuring 109 × 19ft and standing 29ft above the sands. Unfortunately, that same year saw the pier suffering damage below the waterline when it was struck by the Selsey Lifeboat.

The Second World War saw the pier breached for defence purposes, and the shore-end buildings were used as a Royal Naval observation station, titled HMS *Patricia*. The gap was repaired after the war by Mr H. Buxton of the Gaiety Theatre, Manchester, who had taken over the running of the pier. A miniature railway was laid along the pier deck by Charlie

Kunz and a replica LNER 462 Pacific locomotive was used to pull the carriages that could seat up to forty persons. The old pier pavilion on the pier head was used as an amusement arcade and children's playground until November 1964 when a storm severed it from the remainder of the pier. Five months later, on the night of 3-4 March 1965, the pavilion fell into the sea during a blizzard.

In 1966 the pier was acquired by the British Novelty American Pier Co. However, three years later they hinted that the pier may have to be demolished. The pier changed hands again in 1976 when it was acquired by Harrisons (Automatics) Ltd, a local family concern. The pier became a Grade II listed structure, despite the sea-end of the pier being closed off as it was unsafe. In August 1991 the last remnants of the landing stage was removed.

A charitable trust was set up with the aim of securing Lottery money to restore the pier, but their application was turned down in 1998. There was further woe for the pier in the following year when, on 24 October 1999, the dilapidated sea-end was wrecked during a storm and completely demolished. In March 2008 a further 60ft was removed as unsafe, leaving the pier just 350ft in length.

The current owners of the pier are Bognor Pier Leisure Ltd, led by John Ayers, who acquired it in 1996. The shore-end buildings house an amusement arcade, nightclub (opened in 1984 in the former Pier Theatre), snooker club (opened in 1979 in the former cinema) and bar. Every summer from 1979 to 2007 the famous Birdman rally was held on the pier, and attracted competitors from around the world. The record jump of 89.2m was set by David Bradshaw in 1992, but no one has yet to achieve the 100m needed to claim the £25,000 prize. However, a further shortening of the pier in March 2008 meant the event could no longer be held at Bognor. Very little of the pier neck beyond the shore-end buildings remains, and the future survival of what is left must be in some doubt.

11

SELSEY LIFEBOAT PIER

Due to coastal erosion, Selsey's lifeboat station had to be placed at the end of a 900ft-long pier-like walkway in 1934. This postcard shows the lifeboat pier in the 1950s.

In 1960 a new steel approach gangway, boathouse and reinforced concrete slipway to Selsey lifeboat Pier were opened by the Duke of Richmond and Gordon, at a cost of £75,000. It is seen here on a stormy day in November 2002.

12

PIERS THAT NEVER WERE

Bexhill-on-Sea (1895) – No fewer than eleven pier schemes were proposed for Bexhill between 1895 and 1907, yet no construction work was ever carried out.

The resort of Bexhill-on-Sea was developed from the 1880s by local landowner, the Earl De La Warr. In 1883 he contracted John Webb to build a sea wall and promenade from Galley Hill to the bottom of Sea Road. In part-payment Webb received land south of the London, Brighton & South Coast Railway, from Sea Road to the Polegrove. He developed the land as the Egerton Park Estate, and erected the West Promenade in 1886. Earl De La Warr then developed the land to the east, as a select middle-class resort, from 1894. Through his chairmanship of the urban district council and its select committee, the exclusive De La Warr Parade, with its ornamental gates, was created, as were the Sackville Hotel (1890), Kursaal (1896) and Hotel Metropole (1897). In 1881 the first convalescent home in the town, the Metropolitan Convalescent Institution, was opened, and by 1900 it was home to thirty-six private schools. Between 1891 and 1901 Bexhill's population grew from 5,602 to 12,213. In 1902 the town acquired a more direct railway line to London, and staged Britain's first motor car races along the De La Warr Parade. Bexhill also received its charter of incorporation that year, and the council took over its development, adding the Egerton Park extension (1906), Central Parade (1910) and Colonnade (1911).

The earliest proposal for a pier for Bexhill was submitted by the Bexhill Promenade & Landing Pier Co. in 1895, which planned to build a 1,080ft-long pier on West Parade, opposite Park Avenue and the Egerton Pleasure Gardens. The engineer was Joseph Wall. But the scheme soon floundered.

Bexhill-on-Sea (1896) – Opened on Whit Monday, 1896, the Kursaal concert hall was erected on iron piles on the beach, and was meant to have been the first stage of a pier that was never built. The building was renamed the Pavilion during the First World War, and was demolished in 1936.

Bexhill-on-Sea (1897–98) – The Bexhill Pier & Land Co. was formed in 1897 with a capital of £15,000 (3,000 × £5 shares). They had ambitious plans to develop the Marina with housing and amusement centres, and to build a promenade and landing pier. The company was incorporated on 24 August 1897, and by 17 March 1898 162 shares had been taken up. The pier part of the development was then taken on by an associated company, The Pier Co. Bexhill Ltd, which was incorporated on 12 August 1898 with a capital of £30,000 divided into 6,000 £5 shares. Nevertheless, both the pier and land schemes quickly floundered, and at a meeting of the parent company on 6 December 1900

The opening day of the Bexhill Kursaal concert hall on Whit Monday 1896. The Kursaal was said to have been built as the first stage of a pier that never materialised; one of many such unrealised pier schemes for Bexhill. Renamed the Pavilion during the First World War, the building was demolished in 1936.

The Bognor Regis Millennium Pier proposed in 1997 but rejected by the Millennium Commission.

the proposal was passed to wind it up. The Bexhill Pier & Land Co. was officially dissolved on 17 June 1904.

Bexhill-on-Sea (1899) – An unusual design with two very wide, curving entrance arms, one commencing on West Parade, and the other by the coastguard station; they eventually joined up to form a conventional pier 1,320ft long, which terminated with a large hexagonal pier head and landing stage. The engineer was George Ball.

Bexhill-on-Sea (1899) – A 1,400ft-long pier, designed by R.G. Rogerson, to be situated on West Parade by Park Avenue.

Bexhill-on-Sea (1900) – This pier was proposed by the Bexhill Pier Co. and was to have been situated opposite the end of Sea Road adjoining the Kursaal. The promoters had an authorised capital of £30,000, and engaged the noted pier engineers/contractors Mayoh & Haley. They designed a 2,250ft-long pier, subsequently reduced to 1,515ft, complete with a pavilion in the centre that could seat 1,500 people, and a landing stage. The pavilion would have been almost an exact replica of the one erected by Mayoh & Haley on Great Yarmouth Britannia Pier in 1902. The council, however, opposed the scheme, and it floundered.

Bexhill-on-Sea (1900) – J.H. Blakesley designed a pier at West Parade, opposite Park Road, that would have been 1,400ft long.

Bexhill-on-Sea (1900) – To be sited opposite Eversley Road and the Marine Hotel, on land belonging to Albert Bowler, the pier was 2,500ft long and commenced at the landward end with two arms before they joined to form one promenade deck. A landing stage was to be sited at the end of the pier. The engineer for the project was Durward Brown, who also laid plans for a swimming pool.

Bexhill-on-Sea (Central Pier 1901) – The Bexhill Central Pier Co. was incorporated on 24 April 1901, with a capital of £20,000 (20,000 × £1 shares), under the chairmanship of Alexander Rutland Davey. By 1903 the appointed engineer was Edgar Henriques, and the contractor was the Sir Hiram Maxim Electrical and Engineering Co., whose chairman, Jules De Meray, was a director of the pier company. The proposed pier was 1,200ft in length and 25–40ft in width, terminating in a landing stage, and with a pavilion (seating 1,000) to be placed a short distance from the shore. The pier structure was to consist of cast-iron piles, braced with steel struts and tie bars, supporting steel lattice girders running along the length of the pier. The decking was to be of best pitch pine or larchwood, lined by ornamental wrought-iron railings, with bays at intervals for the erection of kiosks. The total cost of the pier was estimated at £15,500, and the work was to be completed by June 1904. However, by that date the directors had all resigned, and the company crashed. It was officially dissolved by the Board of Trade on 8 August 1905.

Bexhill-on-Sea (1906–07) – Arthur Hornby was engaged in discussions with both the Board of Trade and Admiralty regarding a proposed short pier in the vicinity of the

Kursaal, with a 250ft square platform housing a pavilion. However, no company was officially registered.

Bexhill-on-Sea (1907) – The council drew up plans for a 1,000ft pier in line with Devonshire Road, but failed to pursue the idea.

Bexhill-on-Sea De La Warr Pavilion (1934) – The original design for this iconic modernist building featured a two-level pier with a steel pylon at the sea-end, which may have represented a statue of a diver. The pier idea was dropped due to the cutting of costs.

Bognor Regis Pieramid – A £25 million futuristic pier with a pyramid-shaped building at the shore-end, designed by Mike Jupp, a Bognor cartoonist, from original ideas by Bognor resident Steve Goodheart. The plan was for a 1,000ft-long pier with a 50m-tall pyramid at the landward end in bronze and glass. The interior would have included a National Seaside Heritage Centre and space for conventions, exhibitions and concerts. West Sussex county council and Arun district council both prepared the bid to the millennium commission, but this 1996 'Year of the Pier' scheme was also rejected.

Bognor Regis Centinel Millennium – Following the rejection of the Pieramid scheme, Arun district council and the people of Bognor Regis came up with another new pier idea; this time for a twenty-first-century-styled 'hi-tec' structure, to provide a lasting commemoration of the Millennium, at a cost of £20 million. The pier would have been constructed on the suspension system, for a distance of 100m, and would have featured, on the pier head, a lighting display, restaurant and multi-purpose area for local groups. Sadly for the town, in 1997 the Millennium Commission rejected this plan too.

Brighton Central – This pier was first proposed in 1883, 1,200ft long, it would have faced West Street. The engineer was A. Dowson, who built attractive piers at Redcar, Cleethorpes and St Annes-on-Sea. The scheme was revived again in 1886 with a new design by Richard St George Moore. The pier was never built, but St George Moore went on to build the St Leonards and Brighton Palace Piers, as well as the Brighton & Rottingdean Seashore Electric Railway.

Brighton Casino and Marine Palace – An ambitious plan by James Brunlees in 1908 to erect a short, wide pier opposite West Street. The pier would have housed a marine palace, casino, winter garden, assembly and concert rooms, baths and refreshment rooms. Brunlees had previously erected piers at Southport, Rhyl, Llandudno and Southend-on-Sea.

Eastbourne – A commercial pier was planned in 1907, at Langney Fort on the Crumbles. The pier would have been 1,140ft long, curving away to the west, and would have incorporated a landing stage and 1,000ft wave screen breakwater to the south. The engineer was Alfred Carey.

Hastings Alexandra – This pier was proposed by the Hastings Harbour Co. as a rival to the pier built by the Hastings & St Leonards Pier Co. A provisional order was granted by the Board of Trade in 1865 for a 1,200ft-long pier opposite Warrior Square designed by Sir John Rennie. The building of the pier, however, could not commence until the company had expended £40,000 in the execution of new harbour works authorised by the Hastings Pier & Harbour Order of 1862. The harbour was to have consisted of two piers, 1,200ft apart, enclosing forty acres, but neither it nor the pier was built.

Hove (Cliftonville 1868) – Despite the many attempts to promote a pier for Hove, the resort was always reluctant to embrace one, fearing a pier would encourage the masses to come to this rather select watering place. The earliest proposal for a pier was in 1868, when Joseph Wilson, having previously designed Bognor and Teignmouth piers, drew up plans for a 1,200ft-long pier opposite the Brunswick Cricket Ground, by Mills Terrace. The design was similar to Wilson's graceful piers at Hunstanton (1870) and Westward Ho! (1873).

Hove (Cliftonville 1877) – In 1877 the renowned pier engineer Eugenius Birch drew up plans for a 1,200ft-long pier extending from near Mills Terrace, opposite Fourth Avenue.

Hove (Aldrington 1879) – A pier 1,200ft in length was mooted in 1879 to face the promenade between Walsingham Road and Sackville Gardens.

Hove (1883) – George Gordon Page (designer of Margate's Jetty Extension) drew up plans for a 1,400ft-long pier to be placed opposite St Aubyn's Square.

Hove (1887 and 1891) – A pier was proposed in 1887 to face First Avenue, which was revived in 1891. M.N. Ridley (Folkestone and Shanklin piers) and W.M. Duxbury were the engineers.

Hove (1911, 1913 and 1932) – In 1911 Hove-based engineer and architect Henry Hayne Fox, along with Owen Davies, promoted a pier on Medina Parade, opposite Vallence Gardens. In 1912 the Hove Pier Order officially authorised the transferring of the undertaking to Hayne's Hove Pier, Theatre & Kursaal Co., which was officially incorporated on 9 December 1913 with a capital of £120,000. This was to consist of 100,000 shares of £1 cumulative six per cent preference shares and 80,000 ordinary shares at 5s each. The planned pier was very ambitious: 1,450ft long and built using re-enforced concrete, it was to house a Kursaal accommodating 1,000 persons, with medicinal baths and electrical treatments. A band enclosure was planned for the widened centre and a T-shaped pier head would hold a theatre seating 1,500 and a lower landing stage.

On 18 July 1914 a contract was signed with contractors The Waterloo Engineering Co. of 14 Waterloo Place, Pall Mall, London, who were to float the company and underwrite the capital. They were to be paid £10,000: £6,000 in cash and £4,000 by the allotment of £16,000 fully paid up shares of 5s each. However, upon the outbreak of the First World War, the contractors requested an extension of time, but subsequently lost interest.

The order lapsed, although apparently it was revived again by the Ministry of Transport in 1925. The now re-titled Hove Pier Co. struggled on although it was heavily in debt. In October 1930 Financial Issues Ltd agreed to accept 79,993 ordinary shares and 12,500 cumulative shares in full discharge of all claims against the company. A further act was gained in 1932, when Oliver and Roberts were named as engineers and architects. They too had ambitious plans for the pier: it was to be a short, wide structure of re-enforced concrete, with a large entrance deck housing a theatre and a pier head, with an exhibition and dance hall and a restaurant. Also mooted was a railway line in a tube below deck-level, formed by the inner two of the four longitudinal beams. Robert McAlpine was said to have been given the contract to build the pier at a cost of £320,000.

Needless to say, nothing came of this scheme either, and on 3 November 1939 the company was finally dissolved, after twenty-eight years of fruitlessly trying to get a pier constructed. Perhaps if they had been less ambitious in their plans they might have had more success.

Littlehampton – A 70ft extension of the existing harbour entrance pier was proposed by the Littlehampton Pier Co. in 1894. The company, which was led by Talbot Fortescue Haymer, was incorporated on 3 March 1894 with a capital of £20,000 consisting of 20,000 £1 shares. The pier, designed by M.N. Ridley & Moss Blundell, would have run S.S.E. on a higher level. However, no allotment of shares was made, and on 30 April 1895 the failure of the company was registered. Haymer refused to give up hope, and told the Board of Trade on 30 January 1896 that the company was near to commencing business, yet nine months later, on 15 September, it was officially dissolved.

Peacehaven – The original prospectus of the South Coast, Land & Resort Co.'s 'Anzac-on-Sea', in 1916, envisaged a pier, large pavilion and bandstand, none of which were built.

Seaford (1866) – There was a proposal in 1866 for a '470ft promenade pier, jetty and landing place, with all the proper works, approaches and other conveniences connected therewith, for the embarking and landing of passengers, goods and merchandise, and for other purposes, to the west of the Assembly Rooms.' The pier would have been constructed of iron, and stood 12ft above high water. Two pagoda tollhouses would have fronted the entrance whilst, at the end of the pier, landing steps would have been provided.

Seaford (1893) – In 1893 the Brighton, Worthing & South Coast Steamship Co. applied to the Board of Trade for permission to build a 600ft-long pier with jetty and refreshment rooms opposite Causeway Road by the Esplanade Hotel. The engineer was Alfred Thorne, who was responsible for the construction of a number of piers between 1893 and 1908.

Selsey – The 1887 Selsey Railway & Pier Act included a 300ft-long pier close to Beacon House. There was a further pier proposal in 1907 by Selsey-on-Sea Ltd.

West Worthing – In 1863 the Heene Estate Land Co. acquired land for the development of a new town from William Westbrook Richardson, and sold the southern section to the West Worthing Investment Co. the following year. Between 1865 and 1867 they laid out Heene Terrace, baths, assembly rooms, waterworks and the Heene Hotel (later the West Worthing Hotel, then the Burlington Hotel). A second terrace was also built, and Grand Avenue was laid out.

In 1882 a provisional order was granted for a pier in line with Heene Road. The length of the pier was to be 2,060ft, and it was to have an inclination of 1 in 1.73 from the entrance to the pier head. The principal promoter was Gabriel Samuel Brandon, and the lease of the foreshore was granted to W. W. Richardson. The estimated cost was £20,000, and the engineer engaged was none other than Eugenius Birch. However, whilst working on the project Birch met with an accident to his ankle which led to his leg having to be amputated. Within a short time he died and, apparently, his widow refused to give up his design plans for the West Worthing Pier.

A new pier was designed by Joseph William Wilson (engineer of Bognor, Teignmouth, Hunstanton and Westward Ho! piers), but an application had to be made in 1887 to extend the time to build the pier. One of the reasons given, finance not having been obtained, was the withdrawal of promised funds due to the nearby construction of a sewage outfall.

The money for the pier was never forthcoming, although a sale of the West Worthing Estate in 1895 still showed a proposed pier. A further hotel was nearing completion before money ran out and it was abandoned. To be known as the Metropole, it was left in a derelict state for many years and was christened 'Worthing's White Elephant'. In 1889 a railway station was opened at West Worthing, and in the following year it was incorporated into the borough of Worthing.

BIBLIOGRAPHY

Hastings and St Leonards

National Archives BT297/1895 (Hastings)
National Archives MT10/10 (Hastings)
National Archives MT10/1887 (Hastings)
National Archives BT297/879 (St Leonards)
National Archives MT10/23 (Hastings Alexandra)
East Sussex Record Office QDP383 (St Leonards)
East Sussex Record Office QDP305/1 (Hastings)
East Sussex Record Office QDP523 (Hastings)
East Sussex Record Office QDP731 (Hastings)
East Sussex Record Office QDP487 (St Leonards)
Matthews, Mick, *The Decline of Hastings as a Fashionable Seaside Resort* (Hastings Press, 2006)
Koppel, Gary, and Baron, Mike, *The Story of Hastings Pier* (Hastings Pier Co., 1982)
Thornton, David, *Hastings: A Living History* (The Hastings Publishing Co., 1987)
Merchant, Rex, *Hastings Past* (Phillimore, 1987)
Haines, Pamela, *Hastings in Old Photographs: A Second Selection* (Alan Sutton, 1991)
Elleray, Robert D., *Hastings: A Pictorial History* (Phillimore, 1979)
Barnes, J. Mainwaring, *Burton's St Leonards* (Hastings Museum, 1990)
Wales, Tony, *Hastings: The Archive Photograph Series* (Chalford Publishing, 1998)

Bexhill-on-Sea

National Archives BT31/7555/53839
National Archives BT31/9416/69963
National Archives BT31/8113/58552
National Archives MT10/1049
East Sussex Record Office QDP637
East Sussex Record Office QDP633
East Sussex Record Office QDP616
East Sussex Record Office QDP558
East Sussex Record Office QDP619
East Sussex Record Office QDP635
Bartley, L.J., *The Story of Bexhill* (F.J. Parsons, 1971)
Porter, Julian, *Bexhill-on-Sea: A History* (Phillimore, 2004)

Eastbourne

National Archives MT10/15
National Archives CRES 58/505
National Archives MT10/115
National Archives MT81/384
East Sussex Record Office AMS6501/1
East Sussex Record Office QDP694 (proposed Commercial Pier)
Elleray, Robert D., *Eastbourne: A Pictorial History* (Phillimore, 1978)
Surtees, John, *Eastbourne: A History* (Phillimore, 2002)
Surtees, John, and Taylor, Nicholas R., *Images of England: Eastbourne* (Tempus Publishing, 2005)

Seaford

East Sussex Record Office QDP366
East Sussex Record Office QDP550

Peacehaven

Poplett, Bob, *Peacehaven: A Pictorial History* (Phillimore, 1993)

Brighton

National Archives BT31/11515/88757 (Palace Pier)
National Archives BT31/4785/31677 (Palace Pier)
National Archives MT10/1619A (Brighton & Rottingdean Seashore Electric Railway)
National Archives MT6/1109/2 (Brighton & Rottingdean Seashore Electric Railway)
National Archives AT29/206 (West Pier)
National Archives BT31/805/556c (West Pier)
National Archives CRES58/514 (West Pier)
National Archives MT81/117 (West Pier)
East Sussex Record Office QDP707 (Palace Pier)
East Sussex Record Office QDP528 (West Pier)
East Sussex Record Office QDP487 (Brighton & Rottingdean Seashore Electric Railway)
East Sussex Record Office QDP699 (proposed Casino & Marine Palace Pier)
East Sussex Record Office QDP482 (proposed Central Pier)
Carder, Timothy, *The Encyclopaedia of Brighton* (East Sussex County Libraries, 1990)
Ryman, Ernest, *The Romance of the Old Chain Pier at Brighton* (Dyke Publications, 1996)
Bishop, John George, *The Brighton Chain Pier: In Memoriam* (author, 1897)
Bullock, Albert, and Medcalf, Peter, *Palace Pier: Brighton in Old Photographs* (Alan Sutton, 1999)
Fines, Ken, *A History of Brighton & Hove* (Phillimore, 2002)
Gray, Fred, *Walking on Water: The West Pier Story* (The Brighton West Pier Trust, 1998)
Adland, David, *Brighton's Music Halls* (Baron Birch for Quotes, 1994)
Jackson, Alan A., *Volk's Railway, Brighton: An Illustrated History* (Plateway Press, 1993)
Volk, Conrad, *Magnus Volk of Brighton* (Phillimore, 1971)
Owen, Patricia, *The Development of Brighton as a Resort Town* (Royal Pavilion & Art Gallery/University of Sussex, 1996)

Hove

National Archives BT356/7784
National Archives BT31/32172/132657
National Archives MT10/2008
National Archives MT10/1686
National Archives MT10/4
National Archives CRES37/434
East Sussex Record Office QDP541
East Sussex Record Office QDP713
East Sussex Record Office QDP475
East Sussex Record Office C/C/22/4
East Sussex Record Office QDP433
East Sussex Record Office QDP373
East Sussex Record Office QDP443
Scott, Eddie, *Hove: A Pictorial History* (Phillimore, 1995)
Middleton, Judy, *A History of Hove* (Phillimore, 1979)

Worthing

National Archives BT31/14920
National Archives MT10/1725
National Archives MT10/1260
National Archives MT10/1357
National Archives BT31/501/1982
National Archives CRES37/418
National Archives BT31/5797/40642
National Archives CRES58/529
National Archives MT10/592
National Archives MT10/1299
National Archives MT10/1345
National Archives MT10/1852
National Archives MT10/2059
Blann, Rob, *Edwardian Worthing: An Eventful Era in a Lifeboat Town* (author, 1991)
White, Dr Sally, *Around Worthing in Old Photographs* (Alan Sutton, 1991)
Hare, Chris, *The Archive Photographs Series: Worthing* (Chalford Publishing, 1997)
White, Dr Sally, *Worthing Pier: A History* (Worthing Museum, 1996)
Elleray, D. Robert, *Worthing: A Pictorial History* (Phillimore, 1977)
Kerridge, Ronald, and Standing, Michael, *Worthing: From Saxon Settlement to Seaside Town* (Optimus Books, 2000)
Blann, Rob, *Vintage Worthing: Images of a Lifeboat Town* (author, 2001)
Elleray, D. Robert, *A Millennium Encyclopaedia of Worthing History* (Optimus Books, 1998)
White, Dr Sally, *Worthing Past* (Phillimore, 2007)

West Worthing

National Archives MT10/482

Littlehampton

National Archives BT31/5797/40636

Bognor Regis

National Archives MT10/1172
National Archives MT10/592
National Archives BT31/5797/40642
National Archives MT10/1213
National Archives MT10/1852
National Archives MT10/1299
National Archives MT10/1345
National Archives CRES58/529
National Archives MT10/2059
Cartland, James, *Bygone Bognor* (Phillimore, *c.*1980)
Mills, Vanessa, *Bognor Regis: A Pictorial History* (Phillimore, 1995)
Wells, Paul, and Endacott, Sylvia, *Glimpses of Bognor Regis Pier 1865–1990* (S. Endacott 1990, revised 1998)
Davis, J.B., *The Origins and Descriptions of Bognor or Hothamton* (1807, reprinted 1987)
Young, Gerard, *A History of Bognor Regis* (Phillimore, 1983)
Alford, Michael, *Paradise Rocks: A 1930s Childhood in Bognor and a Little Local History* (Phillimore, 2002)

Selsey

Mee, Frances, *A History of Selsey* (Phillimore, 1998)

General Pier Publications

Piers – The Journal of the National Piers Society (various issues)
Adamson, Simon, *Seaside Piers* (Batsford, 1977)
Mickleburgh, Timothy J., *The Guide to British Piers 2nd Edition* (Piers Information Bureau, 1988)
Bainbridge, Cyril, *Pavilions on the Sea* (Robert Hale, 1986)
Easdown, Martin, and Riding, Richard, *A Guide to Collecting Seaside Pier Postcards* (authors, 2006)
Wood, Chris Foote, *Walking over the Waves: Quintessential British Seaside Piers* (Whittles Publishing, 2008)

Other titles published by The History Press

Memories of the Grand Pier at Weston-Super-Mare
SHARON POOLE

From 1946–2008, one of the country's most successful seaside piers – the Grand Pier in Weston-super-Mare – was owned by one family, the Brenners. This book tells their story in both words and pictures. Arising from the ashes of a devastating fire which destroyed the Edwardian theatre in 1930, the pier became one of Weston-super-Mare's best-loved attractions. Stretching 366m into the Bristol Channel, it even earned a mention in Laurie Lee's book *Cider with Rosie* as 'That white pier shining upon the waves'. The family finally sold the pier in 2008, and just six months later fire destroyed the pavilion. This is the story of those years leading up to the fire, when, against all the odds, as Britons eschewed their homeland resorts in favour of foreign coastlines rich in sun and sand, the Grand Pier not only survived, but thrived.

978-0-7524-4990-6

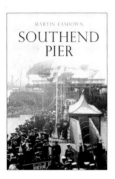

Southend Pier
MARTIN EASDOWN

Southend's Pier was constructed in 1830, and by 1846 was the longest in Europe, at over 7,000ft. Between 1887 and 1890 a new pier was built, and this adopted its predecessor's title of the longest pier in Europe. Millions visited the pier each year, but during 1976 a huge fire engulfed the extensive pier head. With a second fire at the shoreward end in 1977, the pier's future was in danger, but local campaigns to save it have ensured its survival to the present day, despite another fire in 1995, and a ramming by a ship in 1986. This insightful book stands as a monument to this great pier's struggle.

978-0-7524-4215-5

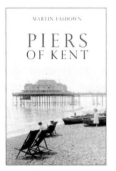

Piers of Kent
MARTIN EASDOWN

With its long coastline and comparatively mild climate, it was no surprise that the County of Kent was at the forefront of the growth of the British seaside resort. From the 1860s a 'mania' developed amongst resorts to erect a showpiece pleasure pier. In Kent, Margate had its first pier in 1855. Even little Pegwell Bay and Tankerton managed to have small piers for a time, yet proposed piers in resorts such as Broadstairs, Westgate, Hythe and Littlestone never saw the light of day. Today, only Gravesend's Town and Royal Terrace, Deal's post-war concrete pier and a stub of Herne Bay survive. This stirring book commemorates what is lost, and celebrates what has been achieved.

978-0-7524-4220-1

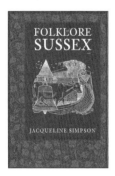

Folklore of Sussex
JAQUELINE SIMPSON

Sussex, though near to London and nowadays extensively urbanised, has a rich heritage of traditional local stories, customs and beliefs. Among many topics explored here are tales linked to landscape features and ancient churches which involve such colourful themes as lost bells, buried treasures, dragons, fairies and the devil. There are also traditions relating to ghosts, graves and gibbets and the strange powers of witches. This book, when it was first published in 1973, was the first to be entirely devoted to Sussex folklore. This new edition contains information collected over the last thirty years, updated accounts of county customs and, alongside the original line drawings and illustrations, photographs and printed ephemera relating to Sussex lore.

978-0-7524-5100-8

Visit our website and discover thousands of other History Press books.

www.thehistorypress.co.uk